The

Spiritual
Universe

The

Spiritual
Universe

One Physicist's Vision
of Spirit, Soul,
Matter, and Self

With a new preface and new illustrations

Originally published in hardcover as
The Spiritual Universe: How Quantum Physics Proves the Existence of the Soul

Fred Alan Wolf, Ph.D.
Moment Point Press

Moment Point Press, Inc.
P. O. Box 4549
Portsmouth, NH 03801

This book was originally published in hardcover as *The Spiritual Universe: How Quantum Physics Proves the Existence of the Soul* by Simon & Schuster, 1996.

Manufactured in the United States of America
This book is printed on acid-free paper

Cover illustration: Orion image courtesy of C. R. O'Dell, S. K. Wong (Rice University) and NASA. The image was created with support to Space Telescope Science Institute, operated by the Association of Universities for Research in Astronomy, Inc., from NASA contract NAS5-26555, and is reproduced here with permission from AURA/ST ScI. Any opinions, findings, or conclusions in this material are solely those of its authors and do not necessarily reflect the views of NASA, AURA, or ST ScI, or their employees.

Cover design by metaglyph
Text set in Adobe Garamond

10 9 8 7 6 5 4 3 2 1

Library of Congress Cataloging-in-Publication Data
Wolf, Fred Alan.
The spiritual universe: one physicist's vision of spirit, soul, matter, and self /
Fred Alan Wolf
p.   cm.
Includes bibliographical references and index.
1. Soul.   2. Quantum theory.   I. Title.
BD421.w65   1999
128'.1—dc20   98-067230
                    CIP

ISBN 0-9661327-1-8

MORE COMMENTS ABOUT
THE SPIRITUAL UNIVERSE

The Spiritual Universe is a comprehensive and provocative examination of the place where religion and philosophy, science and spirituality intersect. Physicist Fred Wolf clearly demonstrates how quantum physics proves the existence of the soul.

Angeles Arrien, Ph.D., cultural anthropologist
and author of *The Four-Fold Way* and *Signs of Life*

This magnificent exploration of the nature of the soul and its place in the universe leaps past conventions of religion and science to a new synthesis, and it is startling: the dichotomy between soul and body is false. What we call the soul is a part of the physical world, but it is a very new definition of the physical. Drawing on traditions of thought as diverse as Aristotle, Chinese medicine, and quantum physics, Fred Alan Wolf paints a lucid, immensely engaging picture of a hitherto concealed world in which the physical does not end with what we see or even with what we can now detect, but swoops on into a vast, profoundly conscious realm where the soul sings songs of God. This is Fred Alan Wolf's great work, a brilliant act of synthesis and insight, a triumph.

Whitley Strieber, author of *Communion*

As usual, Wolf is methodical and clear at explicating physics and thereby provides physics-phobics a wide bridge to understanding some often arcane material.

Booklist

There's a mind-bending place where the mysteries of the soul open to the frontiers of physics. Fred Alan Wolf takes you there in his latest book, *The Spiritual Universe*. From the ancient Egyptian ka to futurist nanotech resurrection, wherever he looks he brings fresh perspectives and dazzling speculations. Wolf is a new Thales for a new physics of the soul; his book will blow your mind and quicken your spirit.

Michael Grosso, Ph.D. author of
The Millennium Myth and *Frontiers of the Soul*

Trendy, but earnest and appealing as well.

Kirkus Reviews

Wolf provides an interesting investigation of the soul—what it is, how it differs from the self, and what role it plays in good and evil. Wolf provides further insights into the "both/and" world of quantum physics as well as the spiritual and scientific basis for the soul. . . . Public libraries would do well to add this to their collections since the discussion invites scientists, believers, and skeptics to a captivating exploration. Both the mainstream religious believer and the New Age participant will find something here to challenge them.

Library Journal

The Spiritual Universe is Fred Alan Wolf's most ambitious attempt to plumb the depths of human existence. . . . In *The Spiritual Universe* he does nothing less than argue that quantum physics is the best perspective from which to evaluate the perennial question of life after death. Wolf proposes that what shamans call "soul loss" is the general malaise of Western society, and proposes the revival of a sacred sense of life to redeem it.

Association for Humanistic Psychology Newsletter

Fred Alan Wolf describes himself as a consulting physicist, writer and lecturer, but the author might also be called a kind of bridge builder. He tries to span two very different—some might say incompatible—worlds. . . .

It is certainly a noble enterprise and has a distinguished literary pedigree, including the works of C. S. Lewis . . .

San Francisco Chronicle

How to Contact the Author

Dr. Wolf travels throughout the United States and the world lecturing on the new physics and consciousness. If you are interested in attending one of these events or would like to inquire about his availability to speak at your event, he can be contacted by mail or email:

Dr. Fred Alan Wolf
c/o Moment Point Press
P.O. Box 4549
Portsmouth, NH 03801

email: fawolf@ix.netcom.com

And check out his web page at www.stardrive.org.

Contents

xiv   Contents

Part 3
The Rising Soul: Into Nothing We Go   133

List of Illustrations

Acknowledgments

I would like to thank the following: Ray De Sylvester, Scott Savage, Bil Thorne and Chris West, Music by Elfheim for permission to use lyrics from "New Age Blues"; Neil Douglas-Klotz for permission to quote his translation of *The Moon and the Sea* taken from the Persian, The Diwan of Shams-i-Tabriz by Mevlana Jelaluddin Rumi, 13th Century Anatolia, in his book *Desert Wisdom;* Sally Potter, Jimmy Somerville, and David Motion for permission to quote lyrics from their song "Coming" from the movie *Orlando,* published by Copyright Control/Virgin Music.

In this paperback edition, I also wish to thank Christus Rex for the right to reproduce a grayscale version of "The School of Athens: Plato and Aristotle," in figure 3.1; Martin Gray (see his sacredsites.com on the web) for the right to reproduce "The Phra Phuttha Chinnarat Buddha" photo in figure 8.1; Gifs Galore for permission to use several graphics, now in the public domain, including Magritte's *Sea* and character faces in figures 7.8, 7.9, 7.10, 7.11, 9.5, 11.1, 11.2, 12.1, and 12.2. All other illustrations are my own.

Thanks are also due to Bob Asahina and Sarah Pinckney for their helpful suggestions and careful editing. I am grateful to Nicki Scully for providing the opportunity for me to travel to Egypt to gather thoughts and research. To my friend and guide, Hakim, much love and thanks for guiding me into the mysteries of ancient Egypt.

To my wife, Sonia, I give my love and gratitude.

Preface to the Paperback Edition

It has been about three years since I sat down to write this book. As many of you know, a lot can happen in three years, and of course much has. While most of *The Spiritual Universe* remains as originally published in the 1996 hardcover edition, I decided to completely update all of the illustrations and include a few new ones. Those of you who are seeing the book for the second time should be delighted by the new illustrations. I believe they convey these difficult ideas even better than in the hardcover edition.

Most of the chapters remain as written earlier. However, I have simplified some of the material, particularly in chapters 7 and 12, which were especially difficult to grasp conceptually. I have also made some minor changes elsewhere throughout the text that I believe will make the book more comprehensible.

The main change, as the new subtitle suggests, is in emphasis. I no longer see this book as a "proof" book. I decided that the notion of quantum physics proving the existence of anything spiritual is problematic at best and impossible to accept at worst. Many readers of the hardcover edition found the idea grand. One reader thought that there were strong parallels to Hegel's thought in works such as *Science of Logic* and *Phenomenology of Mind*. Another enjoyed my descriptions of the work done by physicists, as well as how quantum physics complemented Buddhist beliefs. Many were pleased that I was willing to take on the task of "boldly asserting the importance of 'Soul' in science." But some were put off by the notion of "proof," and thus failed to read the book in the proper light. So the book is not a proof of anything; it is a simple model of one aspect of the spiritual domain of our fragile existence: how consciousness and matter overlap.

How should one read this book? I suggest you read it with an open mind. Keep in mind that science and spirituality, like cultures from opposite ends of the earth, have a difficult time dealing with each other. Words like "proof" have different meanings depending on whether or not you are a scientist. The main reason science is useful is that it attempts to solve our human problems. The main reason spirituality is useful is that it attempts to solve our human problems. It's time that we recognize that the solutions science and spirituality offer are one and the same.

Fred Alan Wolf
San Francisco

PART 1

THE FAULTY GREEK FOUNDATION STONE
FOR A SOUL DEFINITION

WHAT DO WE MEAN by the *soul*? This word, which often enters our lives with such deeply implied meaning, remains a mystery when we attempt to define it. Solving this mystery is the goal of this book! However, while few of us doubt we live in a physical universe, we rarely stop to think we also live in a spiritual universe. How does spirit exist in this universe? Does it fill a volume of space? Does it persist through time? What is this spirit? Is it the same as the soul? Can science help to define spirit and soul? To answer these questions and many more we shall be asking, we need to go back in time and set a foundation for inquiry. In part 1 we do just that. We look at many soulful questions as we ponder our spiritual existence. It helps to know that ancient minds, possibly wiser minds, and deep-thinking minds from the dawn of early Greek civilization attempted to define the soul, the self, the spirit, and their relationship with the physical universe. However, as we shall see, a problem arises as we attempt to follow our Greek forebears in their defining efforts: namely, we may be answering the wrong questions! So let us see how we came to consider the biggest mystery facing all of us—the mortality of individual life and the survival of something so mysterious and yet so much felt by each of us.

CHAPTER 1

Some Soulful but Wrong Questions

My Moon is in Uranus.
My future's lookin' spotty
When I went into a trance,
My soul left me for another body.

—"New Age Blues," lyrics by Ray De Sylvester,
Scott Savage, Bil Thorne, and Chris West; music by Elfheim

THIS IS NOT AN EASY book for a scientist to write. I feel conflicts arising in me as I attempt to put my thoughts into words. These conflicts occur because I should know better than even to attempt to write about the mystery we call the soul. I should know better because I have been trained in that objective information-base of the world called physics—the acknowledged king of the sciences. Unfortunately, this king, unlike Old King Cole, seemingly has no room for any soul, merry, old, or not.

I'm not the first nor the only physicist-philosopher to speculate about the issue of the soul's existence and its seemingly precarious, mysterious, and subtle relationship with the energy and matter of our bodies. As we shall see, Aristotle and Plato also worried about its existence. Aristotle saw the soul as a subtle substance, one that would vanish when the body vanished in much the same way that the sharpness of a knife will vanish when it is melted down in a furnace. Plato, sharing a somewhat similar view—after all, he was Aristotle's mentor—also saw the soul as a substance, but as a nonphysical one, which was eternal, idea-like, and capable of existing beyond the body.

Where does modern science and technology stand in this debate? Can today's physics and computer technology provide us with the hope of eternal life? Set aside these questions for the moment and consider how answers to them might change our lifestyles.

Have We Lost Our Souls to
Modern Technological Life?

We live the good life. Yes, indeed. We are better fed, more protected, and bathed in the light and luxury of countless new technical achievements springing up every day, at least in the Western world. In the so-called Third World countries, the good life of material wealth may be absent, but if all goes as planned by ideal altruistic ruling and governing forces, soon the whole world will enjoy Western-like material wealth.

Many Westerners feel we are approaching utopia: living longer and, perhaps with the aid of science and technology, enjoying more fruitful lives. The subject of life-extension through cryogenic storage (literally freezing the dead) until science reaches the understanding and technology to resurrect the dead is becoming more popular. Although modern medicine promises us longer life and even the prospect of living forever as perhaps programs in a computer or as cryogenically frozen heads, I think few of us take heart from this. Consider the following: Upon resurrection, just what or who would be resurrected?

As we live longer, we face untold population growth reaching into ten billion by mid twenty-first century. Do we have enough material wealth and scientific know-how to support all of these souls? Or should we reconsider whether people are souls at all and, if they are not, should they be subject to the same laws as other plant and animal population controls? In other words, should population control be the right of people everywhere?

This leads us to reconsider what we mean by life and death and what we could mean scientifically by the soul. In the West the question of death is hardly ever considered. Except for immediate and personal tragedy, we see no signs of it anywhere, except for the make-believe body count we watch on TV cop shows and the like and perhaps an occasional news broadcast describing an auto fatality or an assassination of a political figure. Most of us seem to feel we will live forever. Of course many intelligent, sensitive individuals see through this charade.

Beyond the abortion issues, population growth, and frozen dead-heads, there are other darker shadows in the bright light of the setting Western sun and foul-smelling scents in the chilly dusk wind that howls in the future. Our Western approach to life seems to be leading to an ever-growing "cool" isolation—this insularity results in many people finding themselves only able to communicate with the world from behind computer screens or within the confines of an office. We are growing apart from each other, and this lack of communion is taking its toll.

Choking Smokers, Don't You Know the Joker Laughs at You?

Our failure to communicate has a funny side to the chill. My recent trip to New Orleans brought this realization to me. There was a surprisingly cool crispness in the air as my wife and I enjoyed Halloween night in the normally balmy Big Easy. The French Quarter was chock-a-block with the usual ghosts and goblins, but there was something else present—something I would call sinister and funny at the same time.

Many walked Bourbon Street dressed in skull masks. I felt a giddy laughter bubbling inside of me because it appeared that no one was taking death seriously that night. I especially found it amusing to see these walking skullheads blowing cigarette smoke out of their lipless, toothy, grinning mouths.

I felt as if death was present, from the time I arrived in the French Quarter until the moment I left. That made me wonder: Have we in our growing isolation allowed our souls to slip from our grasps as easily as the smoking skulls let loose their clouds of burning tobacco? Is our soul-loss due to our present day isolation?

Machines are of no help in this question. They do what we find to be drudgery. Yet, in ancient days there was wisdom to be gained in the old dictum of Buddhist life—*chop wood, carry water, clean your rice bowl.* Our modern life seems to have made that old dictum vanish into the pages of ancient history. We use machines to chop our wood, to pipe water into our homes, and to wash our rice bowls. Or we go to restaurants where we order from a mechanically smiling waitperson, who, having taken a course in operant-conditioning, responds with a heartless, "Hi, I'm Brian, and I will be serving you this evening. Our specials are. . ."

We need only look around us to become disheartened. People become machines to survive at their jobs. We are ever building labor-saving devices to make life easier for us as we sit in the lap of stupor having nothing better to do with our time than watch *Roseanne* on the boob tube.

The modern world appears to be run by every different kind of machine imaginable. These machines are becoming more and more complex as modern computers grow tinier and tinier. We even see these devices, products of ingenious human thinking, gradually replacing the humans who designed them!

With this growing mechanical disillusionment, something else gnaws at us. Are we *all* soon to be replaced by machines? Artificially intelligent though they be, they're mechanical soulless entities, aren't they? Metaphorically speaking in the hidden-meaning names of rock musical groups from the sixties, our machines and our life styles seem to be leading us to a twenty-first century bleak landscape of modern heavy metal, concrete covered pathways leading to

The Grateful Dead, and lifeless non-floating Led Zeppelins. Today we are facing the notion that we are again a lost generation: a world without soul.

Are we indeed in danger of losing our souls only to be replaced by modern artificial intelligent conveniences? Some scientists[1] believe our souls are nothing but artificial intelligence devices—sophisticated wetware computer programs—nothing more and nothing less. Other scientists believe we will find our souls in the minuscule interactions of atoms and molecules that ultimately fuel the activity of human biological functioning. And to other scientists, possibly like myself, the soul remains a very big mystery not to be confined to the folds of flesh we call our human bodies. Yet, at the same time, is it necessary that it should be found there? Is there someplace else should we look?

Indeed how should I, as a scientist, look for scientific proof of the soul? My physics knowledge is both a gift and a curse insofar as it is needed to define the spiritual universe and its agent, the soul. The gift is that I see, *objectively*, how much of the physical universe works. That perspective gives me a certain peace of mind that the universe is not an accident and that human life is meaningful and purposeful. The curse is that when it comes to seeing essential matters of the heart, *subjectively*, I often see nothing. My scientific mind habitually takes over and I become skeptical and unfeeling. But my path in this life is through my mind as well as through my intuition. So I have to work to gain subjective spiritual insight that is heartfelt as much as most nonscientists may have to work to gain objective scientific knowledge.

Some Scientific Soul-Searching

As a result of this scientific perspective, I have a difficult time blindly accepting what many call "spiritual truth." The sorry state of the impoverished world—often victimized by seemingly false if not evil spiritual beliefs—troubles me. I shudder when I think of the millions of men and women killed throughout the last one thousand years of history because they simply failed to follow the current religious (usually politically based) beliefs of the people surrounding them. I feel somewhat envious of nonscientists who appear to have great spiritual wisdom and a special frustration with my scientific peers who seemingly fail to appreciate the mysteries contained in physics.

Scientists frequently use words such as *soul* and *God* in their book titles (often in big bold letters) to attract readers, but these words are rarely defined, scientifically or otherwise, and the important human issues dealing with these concepts are equally rarely discussed. In fact, one of the major themes of this book is the failure of scientists to ask the right questions—those that lead directly to answers concerning our vital and precarious human condition.

Instead, scientists lead readers down the well-trodden paths of objective inquiry—what I call the wrong questions. Even though these wrong questions are answered correctly, soon enough the reader gets lost in descriptions of neurophysiology and the like.

I can't stress enough the importance of the questions themselves. Accepting answers to the wrong questions can lead to spiritual isolation, a feeling of depression, and to a sense of pointlessness to life and to the existence of the universe.[2] I call this feeling *soul-loss*. I see it as the general malaise of Western civilization—the loss of a sacred sense of life.

But finding the right questions is not easy. The sacred soul does not possess objective qualities in the same sense as a baseball possesses mass or energy. Thus what can science ask about it? I hope to convince you that the soul is just as real as these baseball qualities. In fact, I shall present rational, science-based reasons for its existence in spite of its apparent nonmateriality (and lack of objective qualities) and offer reasons for remembering your soul in everything you do. To prove the soul's existence requires us to find out what the soul is—to come to some agreement on its definition.

Science is not normally interested in nonmaterial, seemingly mysterious, things. At least most scientists do not seem to be concerned with them. I understand why scientists fail to involve themselves with these mysteries. Such things are exceedingly difficult to deal with, and sometimes result in the investigator's giving up previously held shibboleths, particularly those questioning the foundations of science. History has taught us that we painfully clutch our ideas of right and wrong, life and death, good and evil, in order to maintain order in our lives. Yet the result of our clutching often leads to emotionally polarized minds and unfeeling hearts.

Hardly a day goes by when questions concerning the soul's existence do not enter the political, moral, and spiritual arenas. Often, science gets into the spiritual fray. The well-known Scopes trial about the teaching of creation versus evolution in schools comes to mind. The effects of that debate are still felt in our classrooms. No longer can sacred or spiritual matters even be discussed for fear of upsetting parents' closed minds.

Today we again watch as science enters a difficult arena dealing with the creation of life and the maintenance of life-support. Because these are difficult times, people may, with hope in their hearts, turn to science to solve these problems. But, science usually takes the "heart" out of the soul by discussing it so abstractly and so materialistically that we lose the focus of our concern and find ourselves mulling over the wrong questions—the objective inquiries—and even though we may find answers to these questions, they are not the questions we really wanted to ask.

The soul is not an easy subject to deal with either scientifically or spiritually. If I were scientifically ignorant and spiritually wise, this book would be an easier task. But I'm not and it's not. So why do I try? Because a new and original scientific look at the soul is important today. Indeed, the idea of the soul is perhaps the single most significant concept of our time: one that needs a current, scientifically relevant and heart-centered spiritual view.

We need only to turn to today's headlines to see why. For example, abortion is a major concern for our society. During the writing of this book, Planned Parenthood clinics were bombed and shot at by right-to-lifers, while Catholic churches were picketed by pro-choicers. The debate about abortion, the rights of the fetus, and the rights of the mother is not easy to resolve. The issue concerns whether a fetus is a human being and therefore has a soul. The link of the soul with the fetus has not been made by either side. It is as if each side tends to avoid the question of soul presence. Both sides deal with the issue as blind individuals feeling an elephant and drawing different conclusions based on feeling different parts.

If the soul exists, then when does it begin to exist? When does a fetus become a soul? At conception? At three months? At six months? At birth? And, if the soul does not exist, what does come into existence at conception or during these other stages of gestation?

At the other end of the spectrum, issues about the right to prolong life with medical life-support continue. When should a person be taken off life-support? After three days, six days, several months? And what about capital punishment? If we knew what happens to a person at death, would we still condemn prisoners to the death penalty? Suppose that killing a condemned man is shown to produce negative karma and ultimately be the cause of more violence in the future? Suppose we could prove universally that violence begets violence as the Bible says?

Consider children born with lifetime disabilities, some of them without any sign of consciousness. Would we, as a society, feel freer to allow these children to die moments after birth, if we had a perspicacious view of the soul?

Without a new, enlightened scientific view the soul may disappear into the lost pages of propaganda and history, leaving us to wrestle with such issues in the dark. Even worse, suppose we have souls and, because they are often represented as medieval entities, simply do not regard them as real or important. Without a new view we may be in danger of losing our souls, if we haven't already. Worse still, if we continue to ask the wrong questions, putting the soul outside the scientific realm or taking it apart mechanically and without feeling, in spite of finding answers, we may be left hopelessly morally adrift.

The Mystery of the Soul: Can We Find a Scientific Solution?

Some of you might ask, why bother looking for a scientific solution to the mystery of the soul? Isn't science to blame for this soul-searching? Isn't our present soulless malaise—the loss of a sacred sense of life—being caused by science? Isn't science responsible for our present soulless condition? Why, then, should we ask it for answers?

I agree that science, at first glance, may appear to be the worst place to go for answers about the soul. But we shouldn't be too hasty in rejecting it. In our journey to find a more informed basis for the soul, we may find some heartening surprises, provided we look at our findings in a new sciencific and spiritual manner.

You see, science itself is undergoing a major shift in its understanding of matter and mind and is now attempting to deal with what were previously thought to be arcane subjects. As I mentioned, many books are appearing on the search for a scientific basis for God and the human soul. As we explore the science of soul-searching—that is, searching for the soul—we will see that several have taken this path before us with a scientific bent in mind. The list includes Aristotle, Thomas Aquinas, Isaac Newton—but this is not a surprise, no?—and even the ancient Egyptians. Today, you will quickly see by perusing the latest books about the overlap of science, God, and the soul, that most if not all of them attempt either to explain away the soul as a material process, missing its essential points (that it is sacred and immortal) and its essential purpose (that it is necessary for consciousness to exist) or never discuss it at all in spite of the promising book titles.

To make these essential points and to show the soul's essential purpose is the task of this book. I will show what is missing in earlier scientific explanations and once and for all put the soul, as science attempts to put consciousness,[3] into a scientific and spiritual light.

A "New Physics of the Soul"

Until very recently, science concerned itself with defining the universe's attributes as objective processes. Little attempt was made to consider subjective processes as they are. As we near the end of the twentieth century, science again is attempting to define consciousness as a phenomenon emerging from simpler physical processes. The greatest effort seems to be aimed at answering what I consider to be the foundation of all the wrong questions; namely, how does the self-aware entity emerge from deeper and more elementary physical processes? The answer is it doesn't, and that is very difficult to deal with in today's reductionistic science.

My aim is to set up a "new physics of the soul." In it I will show how the soul, the self, matter, and consciousness are, although related, not equivalent. Present science, based on models generated from Aristotle's vision and later developed with the aid of Newtonian mechanics, led us on the wrong reductionistic and materialistic path. It incorrectly reduced the soul and consciousness to purely physical and mechanical energy. At best the soul appeared as an epiphenomenon generated by material processes. When we bring quantum physics into the mix, the error becomes apparent.

Instead we will see the soul as a process involving *consciousness of knowledge*. This process occurs in the vacuum of space in the presence and absence of both matter and energy. From this new vision we shall see why the soul is immortal. This new vision means that the soul had its beginnings when the universe of space, time, and matter first appeared and has its ending when the universe returns to the nothing from whence it came. The major activity of the soul is manifestation of matter and energy and the shaping of the material world by knowledge. Both manifestation of the world and the soul's knowledge of it are tied to quantum physics principles, specifically the observer effect and the uncertainty principle.

The vacuum is fundamentally unstable. Anything that comes into existence did so through the soul's desire to manifest. This desire governs both the appearance of all matter and the relationship of a unified consciousness to matter through the effect of observation spelled out by quantum physics. Thus the soul cannot be seen either materialistically or reductionistically. In fact the soul cannot be seen as a mechanical physical thing at all. The soul's fundamental purpose is the shaping of knowledge into material form.

What Is Interesting and Original About This Book?

In answering the above question I leave contemporary science's search for the material basis of consciousness and self-awareness and offer a new and original concept. I wish to show that *the self is fundamentally an illusion* arising as a reflection of the soul in matter, much as a clear lake at midnight reflects the moon. At the same time, *the soul is not an illusion*, although it is a reflection of spirit. (I'll define what I mean by spirit shortly.) One goal of the book is to show how the concept of self differs vitally from the soul. To do this will require us to venture on a journey of soul-deconstruction and reconstruction, moving backward and forward through time and history.

This tour provides a new vision of an "empty vacuum," and a new realization of how the apparent picture of multitudes of mortal souls is also an illusion and that the "one eternal soul" with "one eternal consciousness" is a fundamental reality. The pre-quantum or Newtonian picture of the vacuum *infinite potential* is simply the non-presence of matter. Long before modern physics, however, a vacuum was seen by ancient philosophers as the potentiality to become anything. It turns out that this ancient view has more in common with the quantum physics view than does the Newtonian mechanical picture. Similarly, our present Western spiritual vision of soul shows that each individual has a soul with a single isolated consciousness. I will endeavor to show that all these nearly countless, separately conscious souls are illusions, reflections of one soul with singular consciousness lasting throughout the span of time our universe persists.

But proving this will not be easy. The relationship between the self and the soul is a mystery and will remain a mystery even if I succeed in explaining it. To understand what I mean, consider the fact that the speed of light is constant for all observers, regardless of how fast they are moving. That is also a mystery, but it has been explained quite well. We know why the speed of light does not change, but when we observe the experimental consequences of it, we are still in awe that nature plays the game it does by allowing space and time to bend to accommodate light's steady speed.

As another example of explainable mysteries, consider that matter does not consist of atomic and subatomic bits of stuff moving in well-behaved patterns as larger chunks of matter apparently do. Matter plays with our knowledge and theories in a game of hide-and-seek we call the uncertainty principle. That there is a principle of uncertainty governing the behavior and existence of matter and energy is a mystery that fills textbooks on modern physics.

We should not forget that in spite of the explanations of science's greatest mysteries, quantum theory and relativity, these remain mysteries. They are awesome and marvelous, because no one knows why the world is made this way. It should be the goal of science to reflect on the mystery of the soul, in much the same way that science considers the mysteries of modern physics and even delights in them.

What Is This Big Mystery of the Soul?

There are many soul-mysteries, as the amusing lyrics at the beginning of the chapter indicate, in spite of the many books written about it and our present-day rationalism. Science in the twentieth century has seemingly not helped us to regain our ancient popular belief concerning our souls: namely, that our souls are immortal and capable of living in one form of paradise or another even after our bodies have long decayed. I suspect that most people believe in an immortal soul in some form.

We who are not scientifically or academically trained may believe in an immortal soul, but then science intervenes and awakens us from our blissful daydream. According to a recent article by philosopher Michael Grosso, a huge split exists between mainstream America and the scientifically educated in belief in the soul and an afterlife.[4] For example, since 1912 behaviorists, in their attempt to establish psychology as an equal science to physics, struck the term *soul* from their dictionary. This admonishment was felt throughout all Western materialistic science. Anything having to do with a soul was suspect and not to be used by serious academic scientists.

Nobel Laureate physicist Francis Crick offered an "astonishing hypothe- sis" that our souls are nothing more than our bodies.[5] According to Crick, the soul somehow arises from the complex interactions between the multitude of neurons making up our nervous systems and brains. I agree that if this is true, it is astonishing.

Think about it. Out of the mud and soil, living souls, like so many plants arising from seeds, pop up. But, if this is true, I feel somewhat dis- heartened, as if I have lost something that will never be returned. It is like be- ing told the truth about the tooth fairy that left quarters in place of my baby teeth under my pillow at night. Can science provide another view besides Crick's? Or does science need to change in order to accommodate what we mean by spiritual truths?

The Difference Between Soul and Self

Whether or not we believe in our souls, we certainly sense that we each pos- sess a unique being inside our skins. We call this entity the *self.* The Moody Blues song, parodying Descartes, says it all: "I think. I think I am. Therefore I am, I think?" Each of us is aware that he or she is aware. We are aware that we all are aware. Even more, we are aware that we are responsible for our lives and even the lives of others. This responsibility that we may call the "essential goodness of humanity" is reflected to us by the incessant knowledge of our own mortality. We know that our actions can heal, harm, enliven, or destroy other life-forms. In brief, we are aware of our souls because we are aware of our impending deaths and our sense of "goodness."

It is this *dual awareness* that concerns me. Unlike other scientists who write about the soul but miss these essential points of the mysteries of death and goodness, I wish to address them directly. Is the soul just another word for the *self*? Or is there something fundamentally different about it? What happens to the self at death if it differs from the soul? Does the soul or the self continue and reincarnate at some future time? More questions! Will science come to our rescue, or are we doomed to face our souls with nothing more than poetic metaphor to soothe us?

The Soul Is a Virtual Process and Not an Entity

I propose a new vision of the soul here, one that explores many of our earlier concepts in light of the tenets of modern science, particularly based as this vision is on the existence of an "intangible, irreducible field of probability"— the quantum physical wave function, from which all physical matter and energy arise.[6]

Many, ranging from modern scientists to perhaps the Buddha, introduce great confusion into the search by not differentiating between spirit, soul, and self. Based on my research, the spirit appears to be virtual vibrations of vacuum energy; the soul turns out to be reflections of those virtual vibrations in time (I'll explain what a virtual process is momentarily); and the self is an illusion arising from reflections of the soul in matter, appearing as the bodily senses as suggested by the Buddha. Hence the three are related but essentially different.

The quantum wave function demonstrates what I mean by a *virtual* process—one that has an effect even though it is not a result in fact. Thus this wave function, although never measured, has extremely important physical consequences. The soul arises alongside this intangible field of probability—as *virtual processes* in the vacuum of space. These processes appear much like reflections of so-called real processes occurring in everyday life. However, these virtual processes have a life of their own, and even though they are never observable themselves, they account for even the simplest things that we do observe.[7]

In other words, the soul is a virtual process and not an entity. Without it, there is no awareness of entity. Here is an analogy: I believe the soul involves us in a manner similar to the way virtual processes involve the ordinary processes of material existence. We know that in quantum physics, virtual processes are extremely important. An example of this is whenever light scatters from atoms or molecules, such as in the everyday occurrence of sunlight scattering from air molecules and producing the blue sky of the heavens. Here virtual electronic processes are involved.

Consider what the electrons in air molecules must do to accommodate this fact. It is fantastic. When light scatters from an atom, each atomic electron excites itself by literally absorbing energy from the light, even more energy than the light contains! Each electron also moves away from that atom in incremental steps from just a few atomic dimensions to an infinite distance all the way across the galaxy! Then each electron makes the long journey back to where it started, again in incremental steps, giving back all of the energy it had absorbed from the light. In the end, the debt of energy is paid back and the light is re-emitted in a different direction from whence it came.[8]

All of this takes place in literally no time at all as a virtual or *imaginary* process. It appears to the outside world that a particle of light has simply scattered from the atom with no change in energy at all and no obvious escape of electrons from the atom. Yet without all of that going on, it is impossible to account for the scattering pattern of light observed when light interacts with any atoms or molecules. In other words, the sky is blue because electrons take virtual journeys to heaven and back! So do we in the process of reincarnation. But I'm getting ahead of the story.

From the Wrong Question to a New Understanding of the Soul

We began our inquiry into the existence of the soul by pointing out, as many of our forebears have done, that the soul is not an easy topic to discuss intelligently. Is the soul material or an illusion? This natural question introduces a gap separating modern science and current spiritual thinking and leads to the split situation we presently find ourselves in. We are led to see material things as real and spiritual things as beyond matter. To find the right trail, we need to retreat to where we lost the scent.

As we move both backward and forward through history these two visions of the soul continually present themselves. At times the soul appears as if it were something quite physical, like an attribute of an object such as its color or its organization. At other times it takes on a deeper, emotional sense, even a feminine form. One is tempted to regard these two visions as scientific (the soul is material) and spiritual (the soul is imaginal), but this turns out to be an error resulting from our asking the wrong question.

The split in visions of the soul started with early Greek civilization. Plato saw the soul as ideal while Aristotle saw it as material. In Plato's *Phaedo*, Socrates clearly characterizes the soul as invisible and yet able to sense the perfection of equality, beauty, goodness and other "perfect" attributes. The material body was seen as imperfect, with fuzzy or imperfect memories, whereas the immaterial soul was seen as perfect and capable of faultless memory. For Plato, the soul was closest to a virtual or imaginal process, while for Aristotle the soul was completely physical and even composed of a fine material, like some form of gossamer.

After considering Aristotle and Plato, we retreat in time to the ancient Egyptians and Chaldeans. This is the place in time when the split doesn't yet exist; the absolute void contains the undivided spiritual and physical universe and provides the origin of all things ethereal and material. Starting there, we deconstruct the old soul and begin to reconstruct a new soul model incorporating quantum physics.

The next step in resolving the conflict between the materialist and spiritualist view of the soul is consideration of the soul's relation with the whole universe. Here we look at the possibility of the soul existing as a computer program at the end of time. This nonphysical model of the soul leads us back to the vacuum where we investigate how the soul could be nonphysical and yet real at the same time. This takes us to the original step in defining the new soul: finding the difference between the self and the soul.

Next, we go to the Buddha's mind concerning the nonexistence of the self and soul. We find the soul not only able to depart from the body, but also from the world of possibility as it disappears altogether, like a magician's illusion. This denial of the soul by the Buddha actually helps explain how the soul can be fooled by itself, and it leads to some original insights into how the soul can become addicted to matter, even polluted by the body!

Then we march forward to modern science's view of the universe. Balancing new with old, we find a scientific view of heaven, hell, immortality, reincarnation and karma. This leads us to see the soul as an essential unified entity despite the large number living upon the earth today. Finally, we learn how the soul speaks to the body and, in the last chapter, how the soul, spirit, self, and matter are all related.

It is my desire that through my attempt to bring the soul concept into the modern scientific age, the old problem of human existence may actually find a solution. From this research and my new model, I believe that I can convince you that although the self disappears at death, the soul continues forever. The real question is how can we bring that awareness into the light so the essential goodness of humanity is continually reflected for all time?

CHAPTER 2

Aristotelian Soul Physics

*What quantities are observable should not be our
choice, but should be given, should be indicated to
us by the theory.*

—Albert Einstein

*In my soul rages a battle without victor.
Between faith without proof and reason without charm.*

—René Sully-Prudhomme, French Poet
La Justice

LET'S LOOK AT JUST WHAT is at stake here. The problem is, honestly, what are we looking for? To determine a scientific basis for the soul we need to consider just what we mean by the concept of "soul." Although it makes good scientific sense to do this, it is a difficult task. We need a well-defined conceptual theory before we can investigate the facts. This much we owe to Einstein, who reminded us that the theory tells us just what we can observe about the universe. So let's look at some theoretical attributes of the soul.

Physics is based on just three concepts: space, time, and matter. So it is natural to ask how the soul could be measured using these three concepts. In other words, let us ask about the soul's material content and its dynamics. For example, does the soul move in space and time? Does it exert pressure? Does it have physical weight or mass? Does it have any other measurable attribute?

As interesting as such questions are, they are misleading. And to see why such questions mislead us, it is important to see how ancient philosophers dealt with the materiality of the soul. As the ancient Greek philosophe Aristotle might have done, here we shall look at the soul's ability to cause action in the body, to move it about. Aristotle, using logic, was the first to attempt to

determine the physics of the soul. Consider, as an exercise in Aristotelian analytical thinking, how we might define the soul using terms from physics. As
we shall see, many metaphors are necessary to outline any concept of the soul.

What Is the Soul?

Most people, if you ask them, would probably say that the soul is not material, not made of substance. Many would suggest that because of this trait it
makes no sense to look for the soul as one looks for a seat of consciousness in
the brain.[1] What are we looking for? If it isn't a material substance, then how
can we find any scientific evidence of it? On the other hand, if physics is based
on time, space, and mass, isn't it a contradiction to look for the soul if it is
nonmaterial? Does scientific investigation ever deal with the reality of invisible or nonmaterial *stuff*? Although to many people it may not seem to, including even some scientists, the answer is yes.

Do You Have a Magnetic Soul?

Consider the simple experiment of sprinkling iron filings on a sheet of paper
beneath which lies a bar magnet. I'm sure you remember the pattern the filings
make on the paper. Although randomly scattered on the paper, they neatly
form into gracefully curved lines, arching from the magnet's north to its south
magnetic pole. This pattern tells us the bar magnet has a magnetic field surrounding it. Is that field massive? Does it have weight? Can we see it? Well, no
light reflects from that spatial field, so if it's there, it's invisible. We need the filings to see its existence. Nor does that field have any weight or mass, although
it moves objects through space or stops them from moving altogether, as if it
were massive. It literally can move a mountain or lift a train above a track. Remember this fact as we consider the soul theories of Aristotle and Aquinas.

But if we didn't know what we were looking for, namely, a magnetic field
in space, we wouldn't have suspected that tiny iron filings would make the presence of the field known. Perhaps we should consider the evidence for the soul in
a similar manner. Where do we look for evidence? What are the iron filings
showing the soul's existence? For Aristotle, the filings are the body itself while
for mystics and spiritual leaders, the answer is in the domain of spiritual values.

Spiritual Evidence? What's That?

What do I mean by spiritual evidence? The soul is often referred to as the spiritual part of a human being, distinct from the physical. As a spiritual part, it
is often believed to survive death.[2] From a moral aspect, a soul can be either

good or evil. Consequently, the soul is subject to happiness or misery not only in its present life but in its life to come. It is even possible for the soul to move from death of the body in the present to the past.

This latter sentence may throw the reader. After all, many of us think the soul moves through time toward the future on an evolutionary path. It would seem contradictory, then, to imagine that the soul would go back in time. But, consider that the soul is ethereal and could move faster than light, as quick as a thought. If the soul is not material,[3] it would certainly be able to do that, for what would slow it down? Given this penchant for speed, and taking into account the modern theory of special relativity that predicts faster-than-light objects[4] moving backward as well as forward through time, we come to the conclusion that the soul, if it is nonmaterial, should be able to move both ways through time. (In part 4, we will consider why a soul would ever want to do that.)

The nonmaterial soul is often thought of as the disembodied spirit of a deceased person. Massless yet contiguous, it maintains integrity, somewhat like a magnetic field. But unlike a static magnetic field, in this view it doesn't vanish when its material source disappears. If you annihilate the bar magnet, the magnet's field also vanishes. Many materialistic thinkers, who also give credence to magnetic fields, believe if the soul exists at all, it exists somewhat like the bar magnet's field. The vanishing of the magnetic field after destruction of the magnet is the same as the extinction of the soul after death of the body.

Perhaps the soul is like a magnetic field, but not as produced from a bar magnet. Instead, I believe it is like the magnetic field arising from a photon[5] or particle of light. Such a field needs no material source for its existence. The mere propagation of the photon through space generates an ever-changing twisted pair of snake-like electric and magnetic fields that continually give and take energy from each other. Perhaps this metaphor also applies to the out-of-the-body soul in its relationship with the body.

The Emotional Soul

The soul is also the emotional part of a human being's nature, the seat of a person's feelings or sentiments. In our everyday dealings we instinctively regard each other as separate souls somehow living inside of bodies. The soul is thus thought of as the essential *irreducible* element or part of ourselves. Some think of the soul as God. (Indeed, the question of the continuity of the soul with God as presented in the idea that all souls return to God is important, and we shall look at this later.) In African-American culture "soul" pertains to music. One refers to another as having "soul" if they have feelings of familiarity or empathy with the culture. In the culture at large, we often sense our souls only when we are in a spiritual environment as felt in worship or heard in gospel music.

The Feminine Soul

The word *soul* arises from the German word *Seele*. According to feminist author Barbara Walker, the word originally arose as a feminine noun (hence *die Seele*) and was used by mystics like Meister Eckhart and Goethe in the same sense as *shakti* in India means the "ultimate feminine reality."[6]

Many of the ancient words for "soul" were also feminine. These include *psyche, pneuma, anima*, and *alma*. God-souls were Goddesses such as Kore, Sophia, Metis, Sapientia, and Juno. Ancients believed every human had a female soul derived from the Mother Goddess through the earthly mother.

Why this emphasis on the femininity of the soul? Perhaps it is a reflection of the nurturing aspect of humanity that is often overlooked, particularly in male-dominated societies. But perhaps something equally scientific and spiritual is implied here. We know all fetuses begin life as females. Male genitalia develop later during gestation. And if we consider certain aspects of ancient Egyptian and Hebrew mystical sources, such as the Egyptian Book of the Dead and the Qabala (also written Kaballah or Cabala), we find that the femininity of the soul has much to do with the feminine principle of structure and creation. Thus the soul's femininity is connected to the feminine principle of magic and creation. (I'll share some greater insight into the sexuality of the soul in chapter 7 where I deal with the question of the meaning of Adam and Eve.)

The World-Soul

There is another sense of the soul we shall touch on here. Rabindranath Tagore once wrote:

> That which oppresses me, is it my soul trying to come out in the
> open, or the soul of the world knocking at my heart for its entrance?

Many mystics consider the soul to be more than a personal Jiminy Cricket inside each individual. They ask us to look out in the world, out in the heavens to see our essential souls. It is not our inner soul that hungers but something outside of us desiring to come in.

Is there more than one soul, or are all souls part of a single world-soul? Maybe the soul inside you and the soul inside me are simply reflections of one soul living in the world of humanity at large—perhaps even in the universe at large. The whole notion of *inside* and *outside* may not apply to the soul as simply as it applies to the body. Certainly our hearts and livers lie within our skins, although there are some exceptions even to this rule. But the soul may not obey this simple dictum. Maybe we ignore the world soul and must continue to reincarnate because we fail to recognize this. If the soul is not confined to spatial or temporal boundaries, then my soul is your soul is the only soul that ever is or ever was or ever will be. If this is true, then what is a soul? (We will return to the accountability of the soul in chapter 12.)

Having briefly explored the spiritual evidence of the soul, including ideas of soul magnetism, the emotional and feminine states of the soul, and the notion of the great world soul, concepts that provide hints to discovering evidence for the soul's existence, let's next look at how the ancient philosopher Aristotle attempted to describe body and soul as a material physical process.

Aristotle: The Soul Is the Principle of Animal Life

Aristotle realized that there is much difficulty in finding soul evidence. Distinct from Plato's idealistic and perhaps mystical view, which we will examine in chapter 3, Aristotle did not separate the soul from the body, and yet he did not equate them. For him, the soul was that by which we live, feel or perceive, move, and understand.

Hard to Answer Questions About the Soul

In *De Anima* (On the Soul), Aristotle attempts to make it clear that any assured knowledge about the soul is one of the most difficult things in the world to obtain. As discussed in chapter 1, the kinds of questions we ask about the soul must be carefully considered. Asking "What is the soul?" requires us to consider how many different ways there are to answer the question. Is there a unique method for answering the question?[7] Does the soul exist as potential or as actual substance? Is the soul divisible or is it without parts?

Perhaps the soul exists not as a plurality of entities but as a composite of parts of one single entity. Should we then look at the functions of the parts of the soul—thinking, sensing, feeling, and so forth—or should we consider what is being thought about, sensed or felt, that is, the action as a cause of these parts arising?

Just as above, Aristotle raised many hard and valid questions in his model of the soul as the principle of animal life. Aristotle pointed out that when one honestly inquires into the subject, particularly with questions about the physicality of the soul, one is necessarily led to the question of the soul's existence.

No Soul Without a Body?

In no case can we see evidence of the soul without involvement of the body. If our souls move us emotionally, then certainly our emotions as well as our senses seem to fully engage our bodies. If our souls are based on our minds, then perhaps thinking is the outstanding exception. It seems not to involve the body; while we think, there appears to be nothing going on. But modern physiology indicates that even thinking involves the movement of signals within the brain. Thus thinking, too, requires a body as a condition of the existence of thought.[8] If there is no case in which we can see the soul distinct from the body, we would conclude that the existence of the soul apart from the body is not possible.

From this argument it sounds as if Aristotle is saying that the soul is the body. But, as he carefully explains, this is not the case. He says there is no evidence of the soul without the body. Before we go into Aristotle's arguments more fully, let us look at how Saint Thomas Aquinas dealt with the issue of the body and soul conflict by basing his thinking on Aristotle's concepts. Then we shall return to Aristotle who believes the motion of the body, as an action of the soul, is evidence of its existence.

Body and Soul As Viewed by a Saint

Thomas Aquinas was a Dominican priest by trade and a medieval philosopher by predisposition. He was born in Italy in 1225. In the *Summa Theologica*, he wrote about the spirituality of the soul rather extensively. Father Aquinas was also quite concerned with defining the physical aspects of the soul. For example, he asked: Is the soul just another form of the body? He carefully reviewed and then refuted both the idea that the soul was nothing more than a form of matter and the idea that the soul was a body by examining the following propositions:

1. The soul is that which moves the body of a human being. Let us take that as given.

2. Anything that moves another body must itself be a body and partake in motion. Bodies by fiat move or "possess"[9] motion. Motion without a

body appears to be unthinkable. Only that which has the ability to move can be called a body and only that which has this ability can move another body. In other words, if it moves another it must itself possess the ability to move and it also must move when it acts on the other body. In this argument he anticipated Isaac Newton's law of action and reaction, which appeared four hundred years later.

The Argument for the Material Soul

Let's look carefully at the two arguments presented by St. Thomas. In Aquinas's time, one first presented the case for the opposition and then the case for the plaintiff. Accordingly, St. Thomas presented the opposing argument that *the soul is nothing more than the body because that which moves a body is another body itself.* Thus in regard to the ability of the soul to animate the body, he said that *nothing can give something that it doesn't already possess.* For example, a fire cannot give heat unless it already has the ability to produce heat. And so a body cannot move unless it already has the ability to move. Thus anything that creates or produces motion must already have the ability or possess the ability to move.

Inert bodies such as rocks can move and therefore possess the ability to move. Accordingly, such bodies must be moved by external sources which are also bodies. One rock falls against another and we have a landslide. But when looking at animals, including ourselves, we attribute that motion to something we do not see directly: our souls. However, the "soul" must be only a different name for the body's ability to move itself. Since the soul is a mover of the body, we conclude therefore that in spite of being invisible, souls are also bodies. Furthermore, between a mover and that which is moved there must be some form of contact. Contact can only take place physically, so if the soul is mover of the physical body, the soul must also be a physical, albeit invisible, body.

The Argument for the Immaterial Soul

Next, Aquinas refuted the above opposition position by presenting an argument that Aristotle gave in volume 8 of his *Physics*, more than one thousand years before. Known as the prime mover argument (which we will examine later in the chapter), it states that not all things capable of moving objects are necessarily themselves physical things occupying space or existing within time as all physical bodies do. In the light of modern physics, we shall be tempted to dismiss this argument based on Aristotelian thinking as faulty. But the argument still makes sense today.

Although Aristotle's model of the known elements (earth, air, fire, and water) and the tendencies of these elements to seek their natural place (which result in the arising of forces and motion of things) has been replaced by Newton's vision of the physical universe, there is, strangely, something very modern in Aristotle's prime mover argument regarding the nature of motion itself. According to Albert Einstein's theory of relativity and our present concept of fields, motion, and the causes of motion of a body are not necessarily driven by contact with other bodies. Something can possibly move without being aided by another physical body touching it, and motion can even take place when there is no other body in contact with it.[10] Thus if the soul is taken as the prime mover of the body, we could conclude that the soul itself is not also a body.

Body and Soul As Viewed by Aristotle

Using a form of debate popular during that day, we saw how Aquinas first set up the argument that the soul was a body because it was impossible for a physical object to be moved by anything but another physical object. Then, Aquinas offered his Christian view which refuted this, appealing to Aristotle's old allegation, the prime mover argument, that stated it was possible for a nonphysical thing to move a physical thing.

The problem we address here is the physicality of the soul. Although Aristotle's argument is based on the notion of an immovable mover, it hints there is more to the universe than matter's presence causing other matter to move. The big question is whether some form of intelligence or willful intent apart from the material body is meaningfully connected to it and capable of separating from it. Or, in other words, can a mind exist without a body? Aristotle argues that the mind cannot exist without the body, but, somewhat paradoxically, that the mind is not the body.[11]

Aristotle makes it clear that the soul is not related to the body in terms of conventional thought. For example, the soul is not just another mass fitting inside the body or an outside mass pushing or pulling on the body. Nor is the soul a spirit imprisoned in the confines of the flesh as if it were inside a cage. It is also not a movement of some thing or pure movement in itself. And it is not measurable by number; the soul cannot be counted nor can it be divided into parts. All parts of the soul, if we think of the soul as if it were divided, would be present in each and every physical part of the body.[12]

The Soul Is a Special Substance

In spite of its seemingly otherworldly qualities, the soul is very real for Aristotle. It has substance, or, better put, it *is* substance of a rather special kind. This substance is real, but it is immovable. It is a base substance, yet it is not ordinary matter, which is subject to movement and to forces. Even though the soul *does not move*,[13] the soul has power to determine and control the actions of the body. Until the soul makes an effort, these actions are only potential: the body remains potentially active and actually inert. After the soul performs, these actions are factual: the body is potentially inert and actually moving. The body for Aristotle is a large mass of *potentia*, nothing without the soul except the potential to take actions dictated by the soul—this otherworldly, nonmoving substance.

Potential and Actual

Aristotle goes further. The soul is the first or primal level of actuality. Things get confusing here as Aristotle develops the ideas of *potential* and *actual.* The soul is *always* active, thus always an actual substance. The body can be both actual and potential. When it is active and therefore actual, it is moved by the soul substance. When it is potential, it is not moving, thinking or doing anything. It is inert or even dead.

Aristotle believed the soul and the body interrelated in a complex manner involving overlapping levels of potentiality and actuality. We might draw them as in figure 2.1.

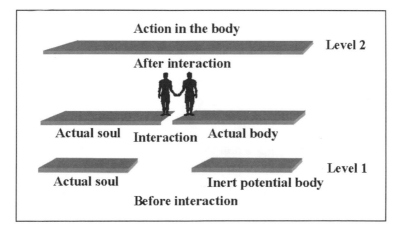

Figure 2.1. Soul Levels. According to Aristotle, what is actually real and potentially real overlap to move the body.

At the primal level (level 1) we find the actual, but immovable, soul and the potential, but movable, inert body. When the soul interacts with the body, the primary levels overlap and the body becomes active (level 2).

Although this may look mysterious, it really isn't when you compare the body and soul to a pile of iron filings and a magnetic field. The field is analogous to the soul and the filings to the body. Just as the filings remain inert unless acted on by the magnetic field, so does the body behave in the same manner unless acted on by the soul.

Aquinas and Aristotle: A Look at the Prime Mover

Aristotle believed the soul was a special immovable substance. One might get the impression that even though the soul was not an ordinary physical body, it should still be able to move, but next we see how Aristotle with his *prime mover* argument points out that the notion of soul movement is misleading and false!

Aristotle's Reasoning: The A, B, C, and D of It All

We need to look at a fashion of argument that was presented in Aristotle's day as pure reason. Today we call it *reductio ad absurdum*, a form of inductive reasoning which shows that a statement one makes specifically as the premise of an argument leads to a general conclusion in contradiction to it, that is, a reduction to absurdity. Using pure reasoning to prove that the soul is an immovable body, the argument goes as follows: Assume the soul to be an ordinary moving body capable of causing movement of any body it contacts. The moving soul-body bumps into another and nudges it. It then causes movement in another body which in turn causes movement in yet another. The question arises: What caused the soul-body to move in the first place? The inductive reasoning argument would state there had to be a prior moving body nudging the soul-body. We would conclude generally from this that a soul moves because it was acted on by another moving body. This turns out to be absurd according to Aristotle.

Imagine a line of toppling dominoes. We see domino *A* tumble. We see that domino *A* fell because it was hit by domino *B*. What caused *B* to tumble? We reason that there must have been another tumbling domino *C* before it and still another *D* before it. This chain of tumbling dominoes *A, B, C, D, E* must continue backward through space and time to the very first tumbling domino, *X*. But then what caused *X* to tumble if it was the very first? Clearly

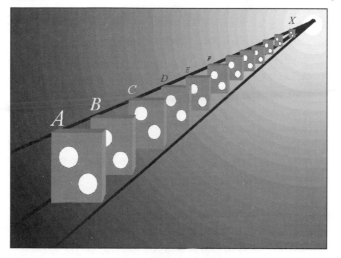

Figure 2.2. *X* Is the Prime Mover at the End of the Line of Dominoes. *X*, without moving, causes all the other dominoes to topple in turn.

it must be a special kind of domino: namely, one that cannot and does not tumble! For if *X* tumbled then we would have to reason there was before it another domino *Y* that was the tumbling cause of *X*'s motion. But we have already stated that *X* is the primal body and therefore one that cannot be moving. Hence at *X* the buck stops. *X* marks the spot of the immovable soul.

It is a fundamental law of Aristotelian physics that if something is moving, something else must have caused that motion to occur. Aristotle merely refutes the notion that the soul must be an ordinary moving body since it is taken to act causally in the motion of a living person. His point is that something can cause motion in a body without itself being an ordinary body. I believe this was also Aquinas's point, although he made it one thousand years after Aristotle. He stressed that the soul is not something that moves as bodies are observed to do, yet the soul must cause the body's motion! This leads to a second point . . .

Does Aristotelian Physics Prove That the Soul Exists?

Aristotle and Aquinas both argued that the soul exists and is the prime mover of every living thing without having to move in and of itself. Aristotle's reasoning was based on his own form of logic: Aristotelian logic—the forefather of modern logic.

The second logical point of Aristotle's argument depends on how you use that term *existence*. We might imagine his argument:

> Consider an ax. If it had no sharpness and no longer was able to cut wood, would you say it still had its essential quality of being an ax? No, of course not. Sharpness is the ax's essential whatness, and being sharp is essential to an ax's character as an ax.
>
> Consider an eye. If it had not sight[14] would it be an eye? No, of course not. It could be a false eye or one drawn in a painting, but that would be an eye in name alone and not in the eye's essential *whatness*.

In this Aristotle creates a metaphor. The word *soul* needs to be used in conjunction with a function or an action. The key here is *entity* or the *inseparability* of a thing from its function and action. By analogy, if the eye were a separate organism, then sight would be the soul of the eye. Similarly, if an ax were a separate living being, then sharpness would be its soul. Thus we cannot separate sharpness from the essential character of the ax nor sight from the essential character of the eye, nor the soul from the essential character of the living being.

The Soul Is the Prime Mover

From this, Aristotle concluded that the soul is the initial actuality of a natural body endowed with the capacity of life.[15] A body that enjoys the capacity of life must possess organs of life. An organ is any functional part of a living body. Thus it is superfluous to ask if the soul and the body are identical in the same sense that it is superfluous to ask if a painting and the canvas it is painted on are the same thing or to ask if a sculpture and the marble it is sculpted from are the same thing. They are in different categories.

The key here is that the canvas is real but the painting is its soul. The essence of the painting is not the physical substance of the paint, but very possibly the information the painting conveys to the eye of the beholder. Thus soul is more like a quality than a quantity. For Aristotle the role of the soul is to move the body in the same sense that the role of the painting is to enliven the canvas.

Aquinas: Movement Without a Physical Body?

Aristotle says a thing like the soul cannot be separated from its function—the body—and that form without function is meaningless. He concluded that since the soul is the initial actuality of a natural body endowed with the capacity of life, and body and soul are inseparable—one being defined by the other as an ax is defined by its sharpness and an eye by its sight—then it is superfluous to ask if the soul and the body are identical. They are not, in the same sense that sharpness and the ax are not identical. Yet the soul is as real as the body in the same sense as sharpness is as real as the ax.

Aristotle has thus established that something else, besides mass, is real. This is a substance that is itself immovable but nevertheless in existence. Aristotle therefore provides the physics for Aquinas to make his nonphysical pitch about the soul's existence. It is important to recognize how influential Aristotelian physics was on Aquinas. It is also important to recognize that our concept of the soul hasn't changed very much since Aquinas made that pitch nearly seven hundred years ago, based on the physics he believed in, a physics that was already more than one thousand years older.

Thomas Aquinas agreed to some extent that the soul was not a physical entity. Aquinas used Aristotle's argument that the soul did not require physical movement. Remember that Aristotle's argument was that the soul, even though it moves the body, is not a body itself and therefore does not move. Hence the soul and the body are clearly not the same thing. For him the soul was the principle of life, feeling, thought, and action in human beings and was a distinct entity separate from the body. In brief, the soul was the essential action[16] of the body; it was the first principle of life in the body but not the physical body itself.

The Real Soul Without Space, Time, and Matter

Now we can see why the notion of a soul is so difficult to deal with intelligently. The prejudice that only things that occupy space are real is an ancient one. If Aquinas were alive today, I am sure he would argue that the concept that only physical things are real is a vestigial remnant much like our appendices, and it no longer serves a real function. Father Augustine, who thought very deeply about the soul and the nature of time and space, had also stated that the soul cannot occupy space as does the body. But the concept that a soul does not occupy space does not dismiss the soul as unreal.

Thus for Aquinas and Aristotle the soul is a real entity even though it does not exist in space and time, although it might be troublesome to consider it as such. For example, it may seem there is little difference between the relationship of sight to the eye, sharpness to the ax, and the soul to the body as presented by Aquinas and Aristotle. What would an ax be without sharpness or any eye without sight?

Yet, although an ax exists in space and time, can we say the same thing for its sharpness? Although an eye exists in space and time, can we say its sight does? Certainly one could say sharpness was real for things like axes but not for tomatoes unless you consider the word *sharp* to pertain to a tomato's acidity; one would not say an ax's sharpness survives the destruction of the ax, nor would one say an eye's sight exists outside of the eye. But what about sharpness and sight? Do they survive the destruction of all eyes and all axes?[17]

Does the relationship of these characteristics of real objects to the objects connect to the association between soul and body? Again you might emphasize that it appears Aristotle and Aquinas are saying the soul is nothing more than an attribute of the person, a characteristic of a living body, albeit an important one, but nothing more than that. One might get the impression that soul is to the body as sharpness is to the ax. There is some element of truth to this, perhaps more for Aristotle than Aquinas, yet we are endeavoring to define here something even more subtle.

Do-Be-Do-Be-Do: Action and Existence

For both Aristotle and Saint Thomas Aquinas, there must be an ultimate source of movement without a physical body. This source being something else besides mass, is equally real. Like the sharpness of an ax or the sight of an eye, there must be a real soul that can exist without space, time, and matter.

What is the difference between an object in itself and what we sense about the object, its qualities or attributes? The very attributes we are talking about—sight in the eye, sharpness in the ax, or possibly, soul in the body— are based on the concepts of being and action, or noun and verb.

Being and action are the two tendencies shown by all things. All things tend to exist or not exist and all things take or give some form of action that we sense. When we look at a thing we tend to label it. *That* is a book or a brook. We label the object as a whole, as a thing in and of itself. However, we base our label on our observations. We see the book has or has not a cover. The cover is red or it isn't. The total of our observations (and I should mention that an observation is more than just a visual appraisal of the object) culminates in a single label: book. Thus, we actually base our experiences of solidity and massiveness on observations of our sense impressions. These sense impressions are fleeting and momentary. They constitute actions of our brains, bodies, and nervous systems.

All and Nothing at All

In a certain sense, therefore, all things are based on nothings or, if you wish, there is no such thing as a noun, because all nouns are based on the actions of verbs.

Before our modern times, and based on the remarkable but nevertheless mystical vision of Isaac Newton, materialist philosophers considered that anything not made of mass did not exist. Little did they know that even the idea of mass was an invention of Newton's, requiring an action even to appreciate its fundamental quality, inertia. For example, pick up an object in front of you. How do you know it is a real object and not just an illusion of your faulty senses? You know the object is real when you pick up the object and feel its inertia, its resistance to movement. The more it resists your attempts to move it, to exert a force on it, the heavier you would say it is. Yet, according to materialistic thinking this quality of action or movement, this inertia, would be an attribute of the mass's existence, but not a substance because inertia has no actual matter. Is inertia real? Most physicists would say that inertia is quite real.

More than likely, this materialistic idea of objective reality became a cornerstone of modern philosophy after the Newtonian revolution. Newton and his forebears set the stage, perhaps unwittingly, for a materialistic and objective view of reality. Accordingly, concepts themselves were viewed in the metaphor of solidity or substance. Questions such as "Does this argument hold water?" or "Does it have any solid basis?" occurred frequently, not only in the laboratories of science, but also in the decision-making organs of jurisprudence and business.

If an object or an argument was suspect, it would be referred to as "flighty" or "fluffy," indicating that "it had no weight." Indeed even today we use the metaphor of mass or solidity to signify reality to such a great extent that few if any other metaphors are even available.

Going back even further to ancient Greek philosophers, we find the same predicament. They were concerned with *being* and *change* as differing concepts of existence. Some would argue for being as the only basis for reality and others would argue for change as the only "thing" that was real.[18]

Nouns and Verbs: Things and Actions, Which Is Real?

If we look carefully at the basis for these arguments describing the reality of objects or their attributes, we find that even though we believe we live in a world of nouns, we continually deal with our fleeting sensory impressions to maintain this belief.

Consider the rush of sensory impressions that greets us when we come into the presence of night-blooming jasmine on a hot windless night. At first we are not sure what we smell. Perhaps it is only a memory, a thought of a long past romance, or the perfumed essence of a stranger in the night. But as we sniff again and again, we begin to build a picture in our minds. The jasmine is real. Then we encounter the woody vine of the genus *Jasminum* filled with yellow blossoms and together the smell and the sight convince us of the flower's existence.

Only when a sequence of sensory impressions seemingly repeats itself are we content with the notion that what has been sensed is real. We might say noun-reality is constructed from repeated verbal sensation. This notion comes pretty close to the Buddhist notion that all is fleeting, all is an illusion.

Action and existence, while not opposites, nevertheless define each other. It's a realization that in a sense all things and nothing-at-all things are based on the relationship of indolent nouns and active verbs. Thus nouns and verbs—things and actions—lead us to question reality. Perhaps in our questions our language gets in our way. We normally deal with action and existence as if both were real, and we describe these categories in terms of things and their movements.

That still leaves us with the question of the soul. Is the soul a noun or a verb? And in either case what does that tell us about its reality? How should we deal with the soul's apparent reality? Upon what should we base evidence of the soul's presence?

In the next chapter we will explore how Plato attempted to define the soul more than two thousand years ago.

CHAPTER 3

Platonic Soul Physics

The soul is like an eye: when resting upon that on which truth and being shine, the soul perceives and understands, and is radiant with intelligence; but when turned towards the twilight of becoming and perishing, then she has opinion only, and goes blinking about, and is first of one opinion and then of another, and seems to have no intelligence.[1]

—Plato

AS A TEENAGER, NATURALLY INTERESTED in the opposite sex, I was always confounded by girlfriends who insisted they only wanted to have a platonic relationship with me. It sounded nice, but it was a real bummer. It meant they did not want sexual contact with me, and in my day that meant no kissing and no petting. The word *platonic* implied something friendly but pure and innocent, not soiled by physical contact. A platonic relationship with a woman signified the perfection of holy heaven, something my teenage hormones were hardly prepared for. In a word, platonic meant ideal, and, for an adolescent, totally unrealistic.

Plato, Aristotle's mentor, had a view different from his student's idea of a body-based soul. He adopted the notion of the *Ideal Soul,* a soul which somehow existed beyond the material. Aristotle obviously rejected Plato's idealism. Even today Plato is regarded as somewhat of an extreme idealist; his theories and ideas would hardly seem appropriate to our modern search for soul evidence. But we shouldn't be too hasty in our judgment of this ancient Greek philosopher. Plato's idealism turns out to be relevant to our search and even lays the groundwork for a modern vision of the soul. Before I tell you more about Plato, let me tell you a little about ancient Greece and its love affair with *logos.*

Figure 3.1. Plato and Aristotle. We can imagine them arguing about the soul's existence.

The Early Greeks and the Power of Reason

At about the same period the Buddha and his followers were developing their spiritual philosophy, a parallel, also outstanding school of thought emerged. This school was, in a sense, almost totally contrary to Buddhist thought. In fact, this school was the only one that even attempted to solve the problems that plagued humankind then and now by using the power of the intellect: the power of reason alone.[2]

From 600 B.C.E. to around 300 B.C.E., Greek civilization flourished and recognized men (not women[3]) had minds—intellects that could discover ultimate truth by reason and reason alone. Plato's thinking was part of the spirit of this time.

Casting everything they dealt with, even the soul, in logical terms, the Greeks conceptualized a *rational* universe, a concept that meant far more than you might believe. They took the attitude that all of nature was ultimately reasonable and followed a plan. In other words, there was an order to all of nature's movements. Anything that moved followed a pattern, and this pattern was rational—capable of being realized as mathematical law. Indeed the word rational comes from *ratio*, the division of one whole number by another. Thus it was that metaphysical schools formed in early Greece with the task of discovering the mathematical design of nature.

The Pythagoreans: Sacred Numbers in Matter

The first major school to accomplish this was the Pythagorean Institute under the tutelage of Pythagoras (585–500 B.C.E.). The Pythagoreans discovered that a wide variety of diverse phenomena exhibited identical mathematical properties. They observed sacred numbers everywhere in the universe. Number was the *first principle* of the plan of nature. All material objects took shapes that were forms of the numbers making them up. Square objects were made from square numbers and triangular shaped objects were made from triangular numbers. All objects were "aggregates" of other objects (in contrast to the Buddhist aggregates discussed in chapters 8 and 11), units of existence—good old solid stuff. The total of these aggregates formed a geometrical figure. Thus square figures were made from square numbers, such as in figure 3.2. Triangular figures were made from triangular numbers, such as in figure 3.3.

In figures 3.2 and 3.3, the number of geometrical units arises from constructing the geometrical form using an integer as the base. Thus, when we

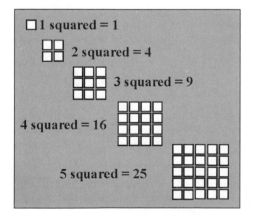

Figure 3.2. Square Numbers. A square number was literally square in form and composed of as many units. Thus 5 squared had 25 units. From this notion we get our present concept of "squaring," meaning to multiply a number by itself.

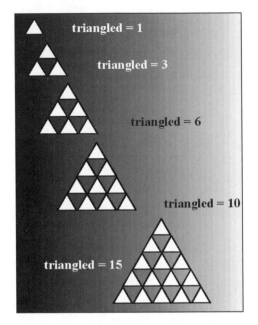

Figure 3.3. Triangular Numbers. Although today we no longer refer to *3 triangled equals 6*, this concept would have been common in Pythagoras's time. A triangled number was literally triangular in form. Thus *6 = 3 triangled,* and *10 = 4 triangled.*

say "3 squared equals 9," we mean the formation of a square figure made up of 9 smaller square units with 3 units on a side (base). Similarly, when we say "4 triangled equals 10," we mean the formation of a triangular figure made up of 10 smaller triangular units with 4 units on a side (base).

Thus, as you can see in figures 3.2 and 3.3, the number one was both a perfect square and a perfect triangle. The number four was two squared, the number nine was three squared, while the number three was two triangled and six was the number three triangled, ten was four triangled, and fifteen was five triangled.

The Pythagoreans believed these numbers were actually contained within matter. In other words, numbers were real particles, made of stuff. A "triangled" number like ten had to be composed of a triangular number of particles. They were triangularly "quantized" in much the same way that quantum physics found that the atomic orbitals of electrons were quantized.[4]

In a sense, the Pythagoreans were the first to discern matter was quantizable, composed of whole numbers of things, and this idea included the form of the material composed of these numbers. For example, if there were 16 particles, they would arrange themselves into a 4 by 4 square form or into 2 triangles: one with 6 smaller triangles (3 triangled) and the other with 10 smaller triangles (4 triangled).[5]

In stringed instrumental music, they discovered sounds were harmonious or not, depending on the comparable lengths of the plucked, vibrating strings. If the strings were equal in length and of the same material, they made the same sound. If one string was twice the length of the other, the sounds were harmonious and made what we today call an octave. Thus a ratio of two to one or 2:1 was a harmonious ratio. 3:2 was also a harmonious ratio making a "fifth." As long as one string was to the other as in the ratio of whole numbers, the sound was pleasing to the ear. In fact, our musical scale derives from the early Greek observation of plucked strings.

Planetary motion was also rational. Planets moved in circles about the earth and therefore had to make sounds, as whirling objects do at the end of a held string. These were the heavenly sounds of the spheres. The sounds they made depended on their distance from the earth.

One, Two, Three, and Four: The Numbers of Heaven

Of special interest were the first four numbers, 1, 2, 3, and 4. They became the numbers of heaven embedded into the mundane and appearing in music, astronomy, geography, and metaphysics. They made up the basis for the whole universe. There were four elements of geometry: the point, the line, the surface, and the solid. There were four elements of existence: earth, air, fire, and water. Of course today we notice the universe exists in a four-dimensional continuum of three space dimensions and one time dimension.

Jungian psychologist Marie-Louise von Franz, in researching parallels between Jungian Depth Psychology and modern physics, devoted a book to the significance of the first four integers as archetypes pointing to the unity of psyche and physis (the psyche and matter).[6] Number one signifies the continuum as a time-bound quality. Two is the rhythm or vibration by which symmetries are engendered. Three is the configuration or spatial extent of the rhythm as actualized in consciousness and matter. Four is the number of closing or wholeness—that which contains the other three.

Similarly, the ancient Hebrews felt the first nine integers were of great significance. In chapter 4 we will look at how this influenced our concept of the soul through the ancient Hebrew form of number mythology called the Qabala.

Idealism and Ancient Realism: The Failure of Our Senses

Following Pythagoras's number mysticism, Leuccipus, Democritus, and Plato carried the idea of rationalism further and also started our Western world onto a path that many feel we have still to recover from. This was the path of *idealism,* separate and quite distinct from *realism.* However, Greek idealism encompassed realism. The idealistic concept of Leuccipus and Democritus was that *real* matter was made of ideal atoms: simple, eternal, infinite in number, different in shape and size, that is, each with definite properties. These properties were objective. They existed as solid little nuts of realism. But we poor mortals were not gifted with such perfect characteristics. In reality, we could only sense these wondrous properties erroneously through our imperfect, out-of-shape senses. These senses always failed to see what is really out there. Thus for them the *real* world was *ideal,* but our senses were inferior and incapable of perceiving it.

With the concepts of Greek idealism and realism holding sway, the failure of our senses soon followed and led to our mistrust of our most human abilities to discern the world around us, a feeling that echoes in our world to this day. It was hopeless to use our senses to discover truth. Truth was objective and out *there* but unfortunately hidden from our inept senses. Humans were despairingly subjective. Only through the ideal numbers was truth to be found, not through our senses. Only through the mind.

Today philosophy takes a slightly different tack. Modern realism states physical objects exist independently of their being perceived (hence no one knows what they really look like), while modern idealism states physical objects consist of ideas, perceived or not. (Objects look like whatever we expect them to, since there are no "real" objects "out there.")

For ancient Greek philosophers, other than Plato, the ideal was "out there," perceived or not, but when perceived, the perception was hopelessly distorted. Plato countered the ancient idealism philosophy of real non-perfect senses coupled to an ideal world, with the notion that any real object was an imperfect *copy* of its ideal. Hence, contained within Plato's idealism, was his realism holding all real objects, perceived or not, as imperfect copies of ideal objects.

With Plato the mind took on great significance as the only organ capable of determining truth. The body could never "see" what was there. The world of ideas was God-like; the world of our senses was not to be trusted.

Plato's vision held that mathematics was the ultimate ideal, the only eternal, the essence of reality itself. If one would only take a few careful glances at the physical world one could "see" what lies beyond—the basic truths of mathematical beauty. If, on the other hand, one becomes too engrossed in the physical, one becomes polluted by it and loses sight of the ideal.

Plato's Idealism and the Soul

When we think of a friend who is over-idealistic and perhaps naïve, but at the same time gifted with an intellect, we may think of the person in terms reminiscent of the Greek philosopher Plato (427 B.C.E.–347 B.C.E.). He came from a family that had long played a prominent role in the social and political scene of Athens, then the center of Greek politics—an often conniving enterprise that disturbed the young Plato. I think of him as an idealistic youth, ever dreaming of distant shores and possibilities and easily disillusioned by reality. He was especially disturbed by the execution of his friend and mentor Socrates in 399 B.C.E. Though it would have been natural for him to follow in his family's footsteps, the intrigues and corruption of Athenian political life so sickened Plato that he declined to do so and instead became a brilliant thinker, writer, philosopher, and the inventor of the concept of idealism, as we know him today.

Let the Philosophers, Not the Poets, Rule!

Plato was indeed as idealistic as we might imagine him to be. His idealism, the death of Socrates, and his belief that our senses fail us in our attempts to describe not only the soul but also anything in the world gave him ample fuel to concern himself with changing politics, ethics, and morals. He strongly mistrusted the political world surrounding him, leading him to proclaim, "Let the philosophers, not the poets, rule!"—a dictum we obviously have not heeded, considering the impact of poets and their emotional appeals in modern advertising.

It was not through politics but through philosophical discourse that Plato sought a cure for society's problems—the works of human minds, corrupt at the core and set on dealing with the lowest principles of human concern. It was in philosophy, Plato believed, that a human's intellect was capable of soaring closest to, not ever touching, the loftiest ideals of the mind of God. He was serious in his hope that not until philosophers became rulers would the problems besetting humankind be resolved and alleviated. To this end Plato wrote voluminously, usually presenting his arguments in the form of intellectual dialogues which, perhaps ironically, were reminiscent of playwrights of his day.

Plato wrote only prose; indeed, he looked with some disdain on poetry and the works of poets and artists in general. This disdain arose from Plato's mistrust of emotions and feelings—centered as they are in the polluted material body and its senses. Truth could only be obtained with a cool and reasoning head—one bent on intellectual pursuit. For Plato, truth was tantamount to God, and all art could do, appealing to the feelings and emotions, was to imitate truth poorly. For Plato, imitation was the lowest sort of the higher human activities. The highest sort was intelligent reasoning. Highest on Plato's ladder of idealism was something not doable by human beings, but only by gods. This included the creation of an original form from which all lesser value copies were made.

To grasp Plato's concept of the soul, and we need to do that in order to understand how modern science can have anything to do with such an idea, we will look at the essence of Plato's thinking about ideals.

Reality, Illusion, and Idealism

Because Plato believed the truth could only be grasped by using reason and not through appealing to emotion, he was particularly disturbed by the worshipful acclaim Greek society gave to Homer's and other poets' writings. If Plato were alive today, I am sure he would be appalled by our society's love for movie heroes and romantic novels. The reason for his disdain for art is easy to understand: Art imitates life, only portraying it through make-believe. It is a representation of reality but not reality itself, and is thereby wanting.

Primary Reality: The Original Given by God

Plato's concern with reality, illusion, and idealism led him to the notion that absolute unspoiled reality only exists at the level of the ideal. Anything not ideal is a copy. Such a philosophy gave us some sense of absolute reality and our failure to achieve it. Instead of living in ultimate and perfect reality or attempting even to reach it, we live secondhand lives accepting imitations of perfection as we are continually jostled by overwhelming media messages—all virtual-reality copies of the ideal.

Plato saw the world in terms of reality, art, and illusion. First was primary reality—the original, created perhaps only once, perhaps by God or lesser gods, and from that point onward mimicked and copied, usually by humans but sometimes by nature, with each copy a somewhat flawed imitation of the original. Copies of primary reality constituted secondary reality—once removed from the source. Next was tertiary reality—often consisting of dramatic performance describing the primary reality, it served no more use than display or advertisement. This third reality was often found in the painting of the primary reality or in the words one wrote or said about it.

Consider thus Plato's description of a bed. In your own bedroom, undoubtedly you will find a bed. Certainly it was built by a craftsperson of some skill. It would be, for Plato, a particular bed and being that today it is certainly more comfortable than it was in his day, it clearly was designed according to some *form* to be as comfortable as possible within the confines of cost, materials, time, and other monetary considerations. Plato directs our attention to the *bed-form*. There can be only one ideal bed-form from which all others are made in mere imitation or in imperfect representation or, in description, flawed depiction. In essence this is how Plato imagined the ideal.

One Ideal, Many Copies

The primary reality—the original given by God—has many copies. As I see it, we distort this ideal through our feelings, a problem that has grown steadily worse in the modern world. Thus, the original bed conceived in the Mind of God was the ideal primary bed reality. The craftsperson's bed would be the first removed from reality, while the picture of the bed or the appearance of the bed in an advertisement would be the second removed from reality. As we see in media advertisements, whether in the newspaper or on television, such ads often appeal to emotions such as sexual desire, comfort, and the appearance of wealth. In this regard, surprisingly little has changed since the days of Plato.

To these three aspects of reality, Plato attributed three "makers": God, who makes the form; the craftsperson, who makes the bed; and the artist, who makes the illusion of the bed. Today, we especially give credence to those makers two steps removed from reality: filmmakers and the advertising industry. Often the advertisement looks much better than the product and the movie's trailer is definitely more interesting than the film.

Plato's Soul Concept: Nothing to Do with Feelings!

Now what does all this have to do with Plato's theory of the soul? The emotional and sensory distortions of humans led Plato to a soul-concept that had nothing to do with feelings and is totally based on *ratio*, meaning rational.

Even today most people, in contrast to Plato's view, believe the soul has the most to do with feelings and emotions and not logic and idealism.

For Plato, the soul was pure, unchanging, simple, invisible, coherent, and eternal. By simple, he meant not complex—made of many parts. Unchanging meant the soul was outside of temporal influence—in fact, eternal. The soul's coherency meant it held together; it was incapable of fragmenting into separate parts. Most important, the soul was rational and, through its rationality, capable of a clear view of reality.

In a recent seminar at the Institute for the Study of Consciousness in Berkeley, California, I asked students what they believed to be the essential difference between the soul, the spirit, and the self. Most replied the soul had the most to do with feelings and emotions, whereas the spirit seemed to imply something at a higher level, and the self was directly known and completely located within the body. This theme that the soul is connected more to our feelings and emotions than to our minds is present in certain forms of music, especially in African-American culture, blues, and country and western music.

Plato's soul, having nothing at all to do with feelings, clearly belonged to a cold, intellectually pure primary reality and as such would only find imperfect imitation in embodiment in the flesh. Even worse, anything we had to say about the soul or its depiction in stories, dramatic representations, or paintings would be much farther away from the real, perhaps impersonal but very pure soul.

Plato's Line and Cave: Reality and Illusion

One of the earliest inquiries on the pure soul comes from Plato's *Republic.* Plato saw the soul as something separate from the body. But, the soul was often confused by being in the body due to misguided thinking brought on by a confusion of the senses. In *Republic*, Plato makes us face our idealistic notions of wisdom and philosophy. He carefully points out that the world of the mind and the world of the senses are divided. For him the world of the mind and the soul is pure, while the world of the body and senses is confused.

In Plato's line and cave models, we see two examples which demonstrate his idealism as it applies to the soul. First I'll describe his line and then his cave. In the line of knowledge and opinion, we find the rational soul's guide to the universe of discourse. Here Plato makes it clear that *ratio* means rational. The creed of the soul is: Knowledge is more important than opinion.

Keeping Your Toes on the Line

Plato's geometrical example of the divided line shown in figure 3.4 illustrates his use of ratio quite well.[7]

The total line A+B+C+D represents everything that is real: everything that can be perceived and thought about. The line has two sides. The upper side shows that which can be perceived: forms, physical things, and shadows and images. The lower side shows that which can be thought about: pure thought, reason, belief, and illusions.

Our perceptions, above the line, consist of *forms* constituting the intelligible realm and *things* constituting the visible realm. The latter are physical things as they are and other things consisting of shadows and images. Our thoughts, below the line, consist of what we know and what we hold *opinions* about. Knowledge is divided into pure thought or insightful intelligence and reason often applied mathematically. In a similar way opinion is divided into belief and illusion.

Plato lays a great deal on what the line connotes to us by its proportions. There is a certain invariant ratio present. For example, the whole line, A+B+C+D, can be divided into two parts, A+B and C+D, and into separate parts, A, B, C, and D. The length A+B is longer than C+D. Similarly, A is longer than B, and C is longer than D. Note the ratio (A+B)/(A+B+C+D). We see that whatever this ratio[8] is (I have taken it to be 3/5), the ratios (A+B)/(A+B+C+D), A/(A+B), and C/(C+D), are all the same value (in my

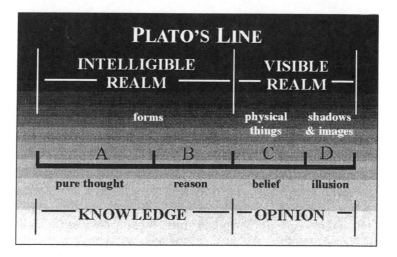

Figure 3.4. The Line of Knowledge and Opinion. The rational soul's guide to the universe.

drawing 3/5) which I call the invariant. The meaning of this invariant is important.

Ratio Means Rational: The Creed of the Soul

In the below-the-line reality, knowledge is more important than opinion as depicted by the lengths of their lines. In the above-the-line reality, that which is intelligible is also more important than that which is just visible in exactly the same ratio and same respective lengths.

There is a direct correspondence between the above- and below-the-line concepts. For example, the things we tend to hold opinions about (C+D) consist of physical things (C), and shadows and images (D). Our opinions (C+D) are broken up into what we believe (C) and what is illusory (D). Illusions are of lesser importance in the opinions we hold compared to our beliefs (D is less than C), and both beliefs and illusions are of lesser importance than what we know (C+D are less than A+B).

Knowledge (A+B) is based on the intelligible realm and consists only of forms. These perceived forms are known to us as pure thought (A) and reason (B). The reader can draw other conclusions from the divided line as are perceivable. For example, the lesser length of the line of opinions when compared to the longer line of knowledge is the same as the lesser length of the line of

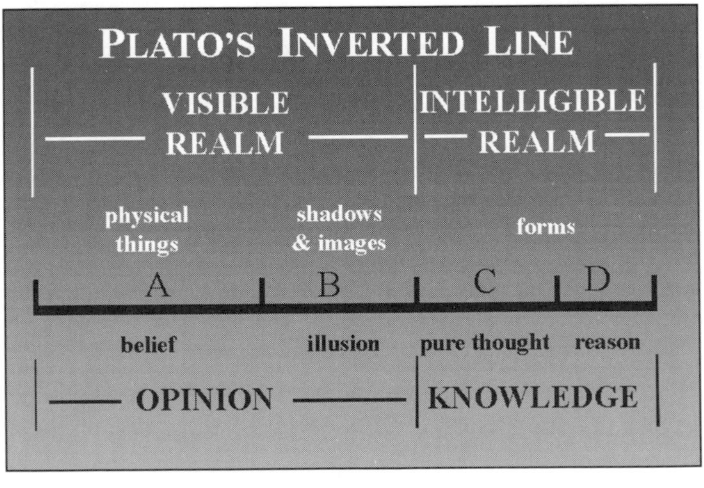

Figure 3.5. The Inverted Line: Our Present Crazed Condition? Souls under this illusion lose touch with reality.

shadows and images when compared to the line of physical things. In all cases, the importance is based on the invariant *ratio* and the relative lengths of the line segments.

When the soul incarnates, it is put into contact with the elements of the line. A wise soul will place the content of its life, lived day-to-day, in perspective according to the invariant rule. It will place no more or less importance on its perceptions than is dictated by the line. This, I believe, is Plato's lesson for the soul.

As an exercise, let me show you what happens if the soul strays from this rule by inverting the lengths and thereby confusion, particularly if the inverted ratios are as illustrated in figure 3.5. Consider this a primary lesson in soul-loss. Step one is confusion by inverting the natural order of the universe.

Here the soul places more importance on opinions than on knowledge and stresses the visible over the intelligible. We might look at this inverted line as the present condition of humanity, caught up in beliefs and illusions and holding firmly to physical things and shadows and images, while relegating ideal forms to lesser importance. Instead of valuing pure thought and reason, the soul considers them to be of less importance than opinions.

As confused as the soul might be in such a condition, it could be worse. Step two is loss of soul by further inverting the natural order of the universe. Consider the following illustration:

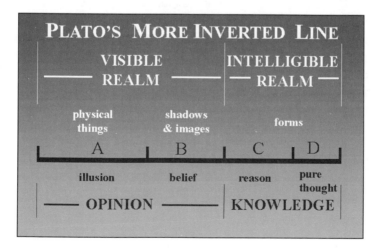

Figure 3.6. The More Inverted Line. Souls following this line are more than crazed: They are lost.

In the visible realm illusions are more important than beliefs. Shadows and images, valued as much as illusions, are more important than physical things, which are valued as much as beliefs. Forms are of lesser importance than shadows, images, and physical things. Reason and logic hold sway over purity of thought or insightful intelligence. Even worse, in this world, what one holds as opinion has far more weight than knowledge. The use of this further inversion played a significant role in Orwell's *1984*. Opinion wiped out the data banks, replacing fact with illusion.

My extension of Plato's line model to include inversions portrays our own society today. It shows how we have turned Plato's idealistic vision upside down, valuing opinion far more than knowledge. A failure to reason, often exhibited by troubled teenagers as well as political leaders who appeal to our emotions over our reason, mars much of our worldview. Whether we have reached the more inverted (figure 3.6) or just the inverted (figure 3.5) position should be considered. I don't know how we feel about the ratio of pure thought to reason today. I do know that it is hardly ever considered, so perhaps it makes no difference. However, I think we still value physical things over shadows and images and hold our beliefs above our illusions, but with the coming of the virtual-reality age, this may soon change.

A Cave-In of the Mind

Plato's cave illustrates the confusion of the soul's embodied senses and the modern dilemma of relying on our senses to determine reality. In this famous scenario, men are chain-bound as prisoners in a cave. They have been in this sorry condition since their childhood. They are so bound that they cannot look in any direction but straight ahead. Outside the cave burns a fire which casts the prisoners' shadows on the wall of the cave. Between the fire and the men a road exists carrying all sorts of commerce—men, animals, carts, and so forth—and yet the prisoners can see only their own shadows and those of the passing travelers. Even the voices of the road travelers, like echoes from a cavern wall, are reflected on the cave wall, so their true origin appears to be the wall. Day after day all the prisoners see is shadows, and eventually they come to believe that the shadows are the only real things in the cave and that they themselves are nonexistent.

Plato's Cave Today: The TV Room

Plato's vision of men so enslaved by chains that they see themselves as only shadows cast on a cave's walls may seem just a whimsical fantasy. But consider our modern cave, the TV room, where, like prisoners bound by the chains of entertainment, we watch the tube, often mistaking the characters we see there for ourselves. Television characters appeal to our baser instincts, as I am sure the shadows did for the prisoners whose own shadows eventually became what they believed was reality. "The medium is the message," Marshall MacLuhan once said, and the people receiving that message are becoming the fictions of the tube. Do you doubt this? Look around at the fashion show of the modern world. Consider the world as made of copies, all seen in the media. Basing the notion of our souls on our feelings has led to modern man's and woman's plight, living through soap opera characters.

In Plato's analogy, when a single man escapes the cave and comes to the light of the sun, he sees the ultimate reality of the world of objects. He is totally dazzled and for a long time believes he has had a mystical vision. When he returns to the cave, how is he to explain his "vision" to his fellow prisoners? Indeed, how are we to explain our own fascination with illusion, gripped as we are by the "cold fire" media?

Why Plato's Cave Is Important to a Scientific Model of the Soul

Plato's parody and my play on Plato illustrate our modern concern with what is essential in life and what is an illusion. To Plato, the shadows are our sensory data, beclouded by the sensorial process itself, and the men bound in the cave are our essential souls, suffering because of our bonds. And as we watch and wait in the darkness of the movie theater or TV room, are we also bound by our sensory appetites for excitement? Are we too beclouded by the sensory process, the process of watching illusion, to ever take part in reality? Perhaps Plato's concept of a cool-headed soul, one unbound by emotional addiction, is just what we need for a science of the soul. Then again, perhaps we are missing something essential by leaving out feelings and emotion. We may need to come to our senses to discover the soul's passion.

Neurophysiological studies indicate our senses are indeed not what they may seem—they are more than means by which we perceive the outside world. It appears that instead of merely perceiving that world, we, like Plato's prisoners, project our ideals and memories into the space-time of our bodies and beyond as we attempt to bring sense data into reason. This leads to the notion of a mechanism by which the soul influences the body and in doing so becomes polluted in the body at the same time. Perhaps this is a clue to the soul's physics. (In chapters 6 and 7 we examine, in the light of modern physics, the way the soul falls into the physical body. The soul's pollution is discussed in chapter 9. Also in chapter 9, I take another look at our senses by following the Buddha's lead.)

Classical Soul Physics:
The Death of Socrates, but Not of His Soul

It is well and good to look for the soul's presence in the live body. Perhaps we take that as given. But we should be equally concerned with the soul's presence at the moment of death. In *Phaedo*, Plato acquaints us with the last hours of Socrates' life. Socrates' belief in the survival of the soul was so strong that he looked forward to his own death with unconcern. Death, says Socrates, is the separation of soul and body, between reality and illusion. By reality Socrates means the soul and where it goes at death. By illusion he means the world of the senses. Socrates tells us again we are not to trust our senses, but are instead to look towards our intellects for absolute justice, beauty, and good—all qualities which are dimly visible to the senses. Thus while we are in our bodies, while our souls are infected with bodily evils, our desire for truth, justice, beauty, and all other superlatives will be continually tainted. The soul, says Socrates, in spite of its bodily corruption, remains immortal and substantial.

With the death of Socrates came the birth of classical Greek soul physics—an attempt to prove logically the existence and survival of the soul. Socrates's belief in the soul and its survival was based on the fundamental ideas of the Greek physics of that time.

Among many who were present, Socrates presents his deathbed arguments to Simmias and Cebes, two followers of the Pythagorean school of physics long admired by Plato. Socrates has to convince them of the soul's immortality and substantiality, basing his arguments on the existing beliefs about and understanding of the physical world: the then-known laws of physical reality.

Doubting Cebes

Cebes expresses doubt about the survival of the soul to the dying Socrates. He urges Socrates to offer his final arguments for its survival, which Socrates does. Socrates asks Cebes to consider: "Do souls of the departed exist in the other world or not?" Cebes agrees. If souls exist in the "land of the dead" after leaving here, surely they could return again to this world and come into being from the dead. But Cebes doubts the existence of this legendary "land of the dead."

Cebes also doubts Socrates' argument for the continuation of the soul after death. Cebes points out that when the "average" person dies, he does so with the belief that his soul is released from the body like smoke from a dying fire's embers—dispersed and destroyed as it passes into the air. Such a soul would simply vanish.

Socrates reassures Cebes that the soul, being substance, cannot vanish, only change form. All substances are indestructible, but their attributes can change. These attributes vary between opposites, one becoming the other. For example, a person is sleeping and then awakens. Another is awake and then becomes drowsy and falls asleep. Sleeping and waking are opposites. Waking comes from sleeping and sleeping from waking.

Cebes agrees. Similarly, if a thing has become larger it must have been smaller to begin with and vice versa. Since living and dead are opposites, and dead comes from the living, it must follow that living comes from the dead. The law of opposites operates in all nature. Since the living must come back from the dead in due course,[9] it follows that the souls of the dead, being of indestructible substance, exist somewhere.

The Soul Survived Death According to Greek Physics

As Socrates faces his death he uses his logical need for idealism and ideal existence as a comfort. Socrates believes the soul *itself* has no opposite. Its form may change but not its substance. With this in mind Socrates faces his own demise with seeming nonchalance. Death—which separates soul and body, takes us back home to the reality of the unpolluted soul and away from the insane-asylum illusion of the body, and frees us of the world of the senses that so confused and polluted the soul—where is thy sting?

By using principles of Greek physics to save the day, but not his life, Socrates posited the indestructibility of the changeable, substantial soul. Greek physics rejected the idea of something coming from nothing and something becoming nothing. If a substance changed, it changed in form only. Substances could change by taking on opposite qualities or by producing opposite effects through involvement in mixtures with other substances (the forerunner of modern chemistry), but they could never vanish.

Opposites and Becoming

What did the Greeks mean by opposites? Normally *opposite* means something literally located across from another object. There must be a space between the objects for them to be opposite each other. They need to be facing away from or toward each other in opposite directions, for example, as the sides of a coin or boxers in a ring. As physical objects, they must be diametrically opposed, contrary to each other: for example, the northeast and southwest corners of a room or the north and south walls of building. But concerning qualities, *opposite* is used as a metaphor. For example, we use the word *opposites* to contrast wet and dry or cold and hot.

Should we also use this word to contrast dead and alive? Here Plato appears paradoxical. What lives and dies? Is he attempting to put the soul in the dead-alive camp? If so, one gets the feeling the soul can die and be reborn.

Relativity of Opposites

The Greek physicists up to but not including Plato thought in terms of absolute measurable opposites, such as wet and dry or light and dark. Plato does not put the soul into this camp. He refers to death and life not as absolute but as relative qualities. Plato wished to modify the absolutist position by introducing the notion of relativity and, most important, the concept of becoming. Things were wetter/drier, colder/hotter, or lighter/darker. One became the other as a result of change of appearance. Thus if a substance became hotter, it had to be its opposite *colder* before. Hence matter could change form but not substance.

The Soul, Being a Nonphysical Substance, Can Never Be Destroyed

Plato's physics held that the soul, although a nonphysical substance, like any physical substance could never be destroyed; it could not vanish or die. Plato argues the soul can change any of its attributes, since these have "opposites generating opposites" possibilities. Thus each one of a pair is capable of generating the other. Just as *wetter* is generative of *drier* and *hotter* is generative of *colder* and vice versa, we are also to consider life generating death and death generating life as changes of form. As sleeping and waking are opposed, so are dying and living. Plato wants us to use this relativistic and progressive physics to deal with the issues of death and life, where no measure or scale is possible.

Form and Substance

Form change? Yes. Complete oblivion? No. Plato's soul was substantial, to be sure. And therefore, although its possible attributes could change by having more or less of a particular attribute or quality, by taking on the characteristics of opposites (an emotional soul or a dispassionate one, a wise soul or a stupid one), the soul could not disappear from the universe or perish. As substance it had to persist in some manner.

Remember Greek physics posed the indestructibility of all substances. From this law came one we are familiar with in modern physics: the *law of conservation of matter*. Until Einstein's theory of relativity appeared, matter was considered inviolate and in fact provided one of the laws of chemical action.[10] With the famous $E = mc^2$ formula, we now know matter (m) can be converted into its opposite: energy (E).

Can the soul be actual substance, perhaps refined, and near to smoke in weight but nevertheless real? I believe that as Einstein's formula has changed our view of energy's being the opposite of matter, quantum physics can change our views about what is real and what is illusionary about the substantiality of the soul. The soul cannot be substance in our commonsense notion, but it is real and substantial in a completely different way.[11] (In part 2 we will explore the soul using new physics concepts and clarify Plato's ideas regarding the soul's substance.)

The Greek Mind-Physics: How the Soul Remembers

Socrates was known for his way of teaching. He believed each person contained all that was known as memory. The teacher had to help the student bring out from memory the truth placed there. Thus, if a question is put to the student in the right way, the student can always answer correctly. Plato calls this the *theory of recollection*. He offered this theory to prove his belief that every person remembers things that are not in his or her immediate experience and that are the recollection of memories of previous lives. Thus, such memories must be evidence of having lived before; ergo, the soul's immortality. For Socrates and Plato, these memories were held by the soul while living in a different body.

But what is memory in the first place and how do the laws of physics explain the workings of memory? Assuming that memory is the *laying down of a recorded track of bits of information*, as in the example of a computer disk or laser disk recording, then when memory is accessed, a probe of some kind must *fetch* that memory for playback. The fetching process requires that the memory record be interfered with.

This is somewhat like the problem of determining the temperature of a substance using a large thermometer. If the thermometer is too big, it affects the reading by adding its heat to the substance being monitored. In quantum physics this problem is magnified considerably, so much so that the mere probing of the substance—a typical example would be the spin of an atomic particle—alters the result completely, often becoming an influencing factor for the outcome.

Why Schrödinger's Student Can't Remember His Past Life

If we assume Plato's belief concerning past lives is correct and that successful Socratic teaching is evidence of an individual's having lived before, why can't everyone remember a past life? The reason depends on the *complementary principle*[12] and the way probing a quantum memory device works.

Attempting to fetch memory is a problem for the fetcher if the process of retrieving memory operates according to quantum physics principles. The observer—the fetching probe—cannot gather the data from memory without altering that data irreversibly. As an example of this, Erwin Schrödinger, the Austrian physicist responsible for the quantum physical wave equation bearing his name, imagined testing an overworked student.

Given that the student has prepared for his examination quite well and is qualified to answer correctly any one of the two questions the professor asks, regardless of which question is put first, one would assume the student would know the answers to both questions. But he invariably answers the first question correctly and then is so disconcerted or tired he entirely misses the second.

One can't assume the student did not know both answers, and yet one can never find out if the student did indeed know them. He could have had all the knowledge, but in the asking of the question, one of those answers was irreparably garbled by the questioner.

Schrödinger's physics student also turns out to be an experimental subject in physiology. Physiologist G. Sperling described exposing several subjects (presumably his students) to a 3 × 3 grid of nine letters or numbers for a fraction of a second.[13] After exposure, the observers typically claimed they could see all of the letters, and insisted they were conscious of all of the elements in the array, yet only recalled three or four of them when asked. If he immediately asked the subjects, after exposure to the letters, to report any three or four randomly cued letters, the subjects answered correctly. The next letters asked for were invariably wrong.

Thus, if you ask the typical student to name the symbols in the first row or the third column or even those running diagonally across the array, he or she can do so, even if the student has no advance knowledge of what part of the array you may ask about. The student is able to recall three or four symbols and then the mind goes blank. It is as if the very action of recalling changes the thing being recalled. This in itself suggests observation of a quantum system.

This example proposes the reason the soul cannot remember past lives. In becoming a physical form, it must follow the quantum rules of matter and energy governing the fetching of memory. Invariably, the fetching process wipes out part of those memories. This process continues throughout the soul's incarnation, from the formation of the fetus—whenever the soul enters the body and a human being comes into existence. Thus to gain knowledge about this life, the soul, in the attempt to put into order what it knows from its past life, inevitably disrupts the pattern of memory. This quantum physics principle keeps us from remembering our past lives and, if I may add, our future ones too.

Plato's Proof of Immortality: Or Why We Have Evil Natures

Contrary to Plato's Ideal Soul, which exists outside of the body and all material cares and woes, the embodied Real Soul can be in danger, although it is immortal. For Plato, the soul is capable of despoilment. It can be fouled by the air, the land, the sea, and even by ugly thoughts. Merely being in a body is enough soul-depredation for Plato. Yet in spite of its befoulment, the soul is immortal simply because moral wickedness, no matter how horrid, cannot destroy it.

If there is no evil that can destroy it, argues Plato, it must persist forever. From this Plato concludes each soul is inviolate, being neither created nor destroyed; each soul has been here for eternity and will exist forever. As he put it in the introduction of *The Immortality of the Soul and the Rewards of Goodness.*[14]

> The soul is immortal because its own specific fault, its moral wickedness, cannot destroy it.

But why does the soul spoil? Why does it have any fault at all? As I see it, for Plato and the early Greeks, the answer was *time*. This is emphasized in Plato's *Timaeus*. Plato's ideal world of perfect *being* exists outside of change, and, therefore, outside of time. This is the natural habitat of the perfect soul. The transitory "real" world of *becoming* betrays that perfection. It is the world of Plato's copies, the world of change and decay. For Plato, this transitory real world is illusory. Hence, once the soul fell into time, its trouble began. While it remained outside of time—and if I may borrow from Einstein for a moment, also outside of space and matter—the soul remained pure essence.

Here Plato tells us quite clearly the difference between good and evil (a subject we shall return to in chapter 9). Good and evil are intimately tied to the immortality of the soul. Evil for Plato is anything that harms or destroys a thing; anything that preserves and benefits a thing is good. That's it. Now evil is rather specific in the manner in which it acts. There are both body and soul evils. For example, in terms of a human being's body, a cold in the head is evil. So are viruses like AIDS. Any disease that affects an organ is evil.

But a soul's specific evil cannot harm the body, and the body's specific evil, in turn, even though it may destroy the body, cannot harm the soul. A soul's specific evil includes primarily moral wickedness and ignorance—none of which kill the body, even though they have persisted throughout history. Although these diseases of the soul can change it, they nevertheless cannot kill it, because the soul is substantial even if it is nonphysical.

Not only human beings are disturbed by evil. Everything in the universe can be upset by it. Take a piece of iron. It rusts, and rust is the iron's evil. A coat of paint that preserves a piece of metal would then be good. All things, alive or not, have their evils. For iron and bronze, it is rust. For wheat, it is mildew. And for human beings, there are many evils—namely, all of the diseases humankind is susceptible to.

Evil has a goal: the complete and utter destruction of the object—the rendering of the object's order into chaos. Thus in terms of modern science, entropy is evil and negentropy (actually the negative of entropy), or the measure of order, is good.

Therefore, we see how our Western culture has so easily come to divide the world and all of its paradoxes into good and evil—so simplistic and yet so powerful.

Take a moment and reflect. The world of wild animal life and its constant struggle with survival makes evil easy to see. When the snake eats the mouse, the snake is evil if we identify with the mouse. And, the snake is good if we identify with the snake. Or, assume you live in the country and you have field mice scampering through your kitchen. You may think your pet cat is good to catch and destroy the mice. If you are a lover of all animal life, however, you might think your cat is evil.

Plato's Argument

In part 11 of the *Republic* Plato argues that all objects may eventually face their own destruction. However, sometimes an apparent evil turns out to be good. Whatever specifically destroys an object is that object's evil. If in spite of an object's arduous journey through many straits and narrows, the object survives, then whatever has happened to it was not caused by evil, even if it looks as if it was. As Plato put it:

A thing's specific evil or flaw is therefore that which destroys it, and nothing else will do so. For what is good is not destructive, nor what is neutral.

Consequently, if anything is found such that its specific evil can mar it, even conquer it or put it through obvious harm but not destroy it, then such an object must be indestructible. Since the soul is affected by evil, in other words, the soul has specific evils associated with it, such as injustice, intolerance, and for Plato, ignorance, we might ask: can any of these evils destroy the soul?

Compare a disease of the body with intolerance of the soul. Clearly the disease if not checked will destroy the body. The body dies from the ravages of the disease. But is it clear that intolerance, for example, kills the soul? It would appear it does no such thing. Evildoers, such as Adolph Hitler, persist in spite of any evil they perpetuate upon others. Some Machiavellian philosophers might argue the evil person grows stronger rather than weaker from evil-doing. But others might argue the soul's specific evil eventually weakens the body. Plato counters this. Suppose we consider a person poisoned by bad food. Suppose the person dies. Could we then argue the food's specific evil—its rottenness or decay due to spoilage or bacterial infection—was the cause of the person's death? No, says Plato. Certainly the food started some form of reaction in the body. That reaction, say stomach hemorrhaging, eventually killed the person. It wasn't the bad food that did him in, it was his own body's response to the bad food that laid him low.

Evil Is Rather Specific

Specific things have specific evils, and only specific evils can affect these things. Just as bad food cannot kill or harm a body[15]—only the body's specific evils can do that—just so a soul's specific evils cannot harm the body or the body's specific evils harm the soul. Granted that a specific evil for one thing can cause a response in another thing, Plato distinguishes the body's and soul's specific evils as essentially different. A body may therefore become quite ill, even losing parts due to illness or accident, and yet none of these specific body evils would cause any harm to the soul. In the same way a person can be a mass murderer, a child molester, a bludgeoner of helpless people, and yet nary suffer any bodily harm. Nor would these monstrous acts kill his soul.

Does wickedness kill the body? Suppose wickedness and immoral behavior could injure the body and cause it to die. Should we then suppose

PART 2

FROM ANCIENT EGYPT AND ALBERT EINSTEIN
TO ADAM AND EVE:
THE FOUNDATION FOR A NEOSOUL

OUR SEARCH CONTINUES. Aristotle taught us that the soul is an essential substance that moves the body. Plato taught us that the soul is an indestructible, nonphysical substance that is pure when out-of-body but polluted by the bodily senses when in contact with them. Who is correct? Or is there something missing—something infinitely more subtle about the relationship between the soul and the body? In attempting to answer the wrong questions and looking for the soul's materiality or lack of it, our journey leads to a blank wall. In part 2 we begin the process of deconstruction of the old soul in order to reconstruct a new one. To accomplish this, we need to move backward and forward in time simultaneously. We must go back before Greek rationality set the stage for the modern soul, returning to prephilosophy time, to Egypt, ancient Chaldea, and the Qabala, for the first look at how nothing becomes something. We will then quantum leap to the present scene set up by Albert Einstein, general relativity and cosmology, and the modern computer view that takes us to the edge of the universe and back. And, then we will go back into time and simultaneously forward into timelessness—into the vast nothingness of vacuum space. Here, there is no space, no time, and yet we find the firmest foundation for the neosoul—a spinning hole in a sea of nothingness. Taking a spin between heaven and earth leads us to all of the problems of self and soul—the crisis of Adam and Eve and a new foundation for the soul: nothing, nothing at all!

CHAPTER 4

The Ancient Basis of a Modern Soul

I am he who came into being, being what I created—the
creator of the creations. . . . After I created my own becoming,
I created many things that came forth from my mouth.
　　　　　　　　　　　—the words of Temu, *The Egyptian History*
　　　　　　　　　　　　　　of the Creation of the Gods and the World

O keep not captive my soul. O keep not ward over my shadow,
but let a way be opened for my soul and for my shadow. . . .
keep ward over the spirits who hold captive the shadows of the
dead who would work evil against me.
　　　　　　　　　　　　　　　—Egyptian Book of the Dead

IN PART 1, AFTER ESTABLISHING the Greek concept of the soul based on the ancient physics of that time, we found ourselves in a paradox. Was the soul substantial, immovable, eternal, and yet still subject to disharmony, wickedness, and suffering? What happens to it when it falls into the matter of the body? Out of the body it is, as Plato and Socrates reminded us, pure and unpolluted, ever sensing its immortality. Why does it suffer so when it falls into the temporal world of the body? If evil obviously can harm so many, why doesn't evil harm the soul that commits it?

The answer to these and other ancient paradoxes must come from a new vision of the soul—one that transcends the ancient Greeks as much as quantum physics transcends Aristotelian physics. In part 2, we begin a search for a new conception of the soul, while at the same time we take apart the old one.

A new soul vision is needed especially today when we perhaps feel ambivalent about the overlap of spirituality and modern technology We are either scientifically sophisticated and, at the same time, spiritually ignorant or vice versa. Concepts like soul, spirit, self, and God, are not fashionable among scientists. Where they are used by scientists they are no more than metaphors for a human being's needs to deal with the immensity of today's problems. On the other hand, concepts like universe, electron, quantum, and space-time, are no more than powerful myths to the nonscientific mind. It knows they somehow exist, but they have little use in solving the real problems of humanity.

Look Backward, Angel!

Why do we need to go back to a time before the Greeks for a new model of the soul? Many years ago a contemporary American composer, whom I met in Israel, told me that today's music, like modern physics, had reached the extreme of abstraction. Composers were using nearly anything to compose on and with, and the science of sound had nearly washed away the delicate creativity and powerful emotional sense of music. Composers were actually considering using the pain threshold of sound as a means of artistic expression instead of the relationship of notes. This composer would have none of this and, in his rejection of modern music, retreated to the study of the grand masters of seventeenth-, eighteenth-, and nineteenth-century musical composition to find a direction for the future of musical composition.

A similar situation exists today in our search for the soul. Some scientists, basing their belief firmly on the current trend of greater abstraction and reductionism, seek the soul in the binding together of emergent physical, neurophysiological, processes.[1] Others, following the trend set by the modern computer, seek the soul in its complex binary programming.[2] While both of these approaches appear promising, I believe that they will eventually lead to blank walls, just as, in its seemingly random assembly of punctuated notes, modern music has lost the meaning of tonal artistic expression.

In a way, the Greeks are responsible for our predicament. They set us on the present course of rationalism. Perhaps we need to realize that both the modern computer vision of the soul and the neurological model rest on Plato's shoulders as assuredly as they do on Aristotle's. Plato told us that our souls were ideal *stuff*, and when we die they continue. We believed him. Aristotle told us our souls were material *stuff*, and when we die, they vanish. We believed him, too. The course of the discussion continues to the present, meandering from one shore of possibility to the opposite, with scientific knowledge seemingly championing both.

On one hand, modern science tells us (see chapter 5) that, at best, our souls are computer programs in the wetware of our brains, and when we die, they will also vanish, or, if we are "lucky," will resurrect as virtual-reality simulations in molecular-sized microchips at the omega point of universal history (the end of time).

On the other hand, modern science tells us that, at worst, our souls simply do not exist at all; they are delusions of grandeur or wishful thinking brought on by our fear of death! Freud told us virtually that early in the twentieth century. And yet there are reports of afterlife and out-of-body experiences and numerous accounts of spirit and ghostlike apparitions, not to mention past-life regressions that are gaining respectable scientific currency.[3]

Early Visions of the Soul

We live in strange times. Perhaps we have always lived in strange times, and perhaps there is nothing really new about facing the problematic existence of the soul. So, like our modern composer seeking new music, it is helpful to go back in time and look at other strange times, in the prephilosophical eras of ancient Chaldeans, Hebrews, and Egyptians, to find some guidance.

This backward-through-time look is imperative because although these people, compared to ourselves, apparently held very different views of nature, the soul, and life, a somewhat similar view to theirs is reappearing today. During the atomic age, shortly after the bombing of Hiroshima, scientists tended to view nature as wild and uncaring, distinct from us in being "out there," and perhaps even dangerous. Our desire for security, warmth, and sustenance often forced modern technology to conquer nature.[4]

This early modern vision of raw and to-be-conquered nature was vastly different from that of the pre-Greeks. This pre-Greek vision of the soul, life, and death is what I see returning to our time, albeit, along with it some strange views too![5] To see what has arisen, we will consider carefully the prephilosophical vision of our forebears, for in it the soul and nature were not separate. Everything in nature, including rocks, mountains, sky, animals, and plants, was alive, bristling with sacred energy.

A Brief Look Backward

According to many historians, our ancient forebears saw God everywhere, in all nature and in all of the universe. Egyptologists date the prephilosophical Egyptian culture as far back as 3150 B.C.E. Sumerian writings found on clay tablets date as far back as 4000 B.C.E. The myth of Isis and Osiris can be dated as far back as 2500 B.C.E. The Egyptian Book of the Dead, documenting beliefs from the eighteenth Egyptian dynasty considered part of the New Kingdom, can be dated around 1500 B.C.E.[6] Abraham, the Chaldean and non-Jewish father of the Jews and Arabs, left Ur, now part of Iraq, just one hundred miles northwest of the Persian Gulf, sometime between 2000 and 1900 B.C.E.[7] In these events and times people just like us were attempting to deal with life, death, happiness, God, and the immortal soul.

God Is Everywhere and in All Nature

Gods were seen as nature itself and observed everywhere. But nowhere else is this vision brighter than in Egypt. Picture a desert, vast and seemingly lifeless. Now imagine a gash of hot, wet, green, teeming life cutting across this brown arid wasteland, and you see a clear demarcation of life and death. Such was the impact of the Nile riverbed upon the ancient mind. Stand with one foot along the bank of the Nile and the other on the hot desert sands, and you feel the life of this life/death presence as certainly as you feel it pulsing in your blood or hear it in your expelled air.

Go outside your shelter at night and look up through the dry night air at the Milky Way galaxy. At once you are transported. That Milky Way *is* the Nile, no denial (pun intended). The terrestrial desert reflects the night sky and Egypt was seen as the mirror reflection of heaven. To that sky all Kings of Egypt shall return.

Egyptian gods, depicted in the Egyptian Book of the Dead, for example, inhabited the *neterworld.* Neter meant *God* in old Egypt. But today we take neterworld to mean the netherworld or the underworld. This is also accurate in a certain sense as we will see. The word *god* is also imprecise in Egyptian religion. Neter actually refers to a spiritual presence or essence. Many of the ancient peoples knew this essence. The Hebrews called it by the Hebrew letter א (aleph) or referred to it as the "breath of God" and symbolized it by the letter ש (sheen). We will look at their conception of this sacred "wind" later.

In the Beginning

I am Atum when I was alone in Nun (the primordial waters).[8]

In the beginning, similarly to our image of the big bang, our ancient forebears envisioned the universe forming from a great void sometimes imagined to be primordial waters. Likewise, they discovered the Creator of the universe was not engendered in any particular place. Instead, He or She was in all of nature and in all of the universe, constantly becoming. Thus we have the paradox of the Creator's becoming and remaining eternal, being contained by the creation He[9] created!

According to the Egyptian Book of the Dead, the sun god, Atum, was completely self-created. Indeed, his name means both *everything* and *nothing*. Atum can also be compared with the Hebrew letter aleph and is sometimes referred to as Temu or simply Tem. Perhaps the Greek word *atom* can also be related to the Egyptian word *Atum*: the primal substance. We see this name as something that is perfect, complete in itself, and thus contains, paradoxically, *everything* and *nothing*—terms that share these qualities of perfection. Atum is always present at the beginning of time and at the end. Atum is a pregnant stillness like the feeling of the air just before a great rainstorm or hurricane. In a real sense Atum is *nothing*—potentially able to be anything.

Atum named and produced eight gods and goddesses and so began to cast form from that which was chaotic and formless. Before the creation of the universe, these eight precreation gods paired off into four couples, each with a god and a goddess. These gods represented aspects of the primal chaos that existed and consisted of the following:

- Nun, representing the primordial waters, and his consort Naunet, the counter-heaven;
- Huh, representing the boundless stretches of primordial formlessness and his consort, Hauhet;
- Kuk, the darkness, and his consort Kauket (yes, we most likely get the word *coquette* from this word);
- and finally, Amon, the hidden or secret holding that is intangible unseen chaos, and his consort, Amaunet.[10]

A slightly different version comes from *The Pyramid Texts*:[11]

> When heaven had not yet come into existence,
> When men had not yet come into existence,
> When gods had not been born,
> When death had not yet come into existence.

Atum spit Shu and sputtered Tefnut from his mouth. In another version, He masturbated them into existence. Here two of the Greek elements are created: Shu, representing air, and Tefnut, moisture or water. From the marriage of air and water both Geb, the earth, and Nut, the sky, are created. These two then mate and produce two couples, Osiris and his consort Isis, and Set and his consort Nephthys.[12]

All of these gods and goddesses arose from the supreme vacuum, Atum, who seemingly reversed entropy and separated air from moisture and then condensed them into sky and earth. Although this is presented as ancient, conceptualized by ancient minds, it is not far different from the big bang model of modern cosmology, wherein something, namely, all that exists, spontaneously burst from the vacuum, reversing entropy like crazy. This conception is the clue to our new model of the soul. *The home of the soul is not the physical world, it is the vacuum world, the world that is able to be potentially anything.* (We will return to this in chapter 6.)

Versions of the creation differ geographically. In old Egypt there were four major religious centers: Heliopolis, Memphis, Hermopolis, and Thebes, and each of them had slightly different versions of the beginning. In Memphis, Ptah and Sekhmet played major roles and were added to the list. In Hermopolis, Thoth, the principle of wisdom and the moon-god, is elevated to the list and Ra, the principle of light, eternity, power, and rebirth attains sun-godlike prominence. In another version, the male god Shu, the dry air, and the female goddess Tefnut, the mist, are the twin children of Ra, the breath of light.[13]

It turns out that Temu and Ptah are actually aspects of one neter, like sides of a coin. Temu is thought of as the primordial act, the first creation, a pure essence and spirit. Ptah is believed to be Temu come to earth, the same principle of spirit but, in this instance, the manifestation of the act of creating matter. Of the great nine neters, the remaining seven are also paired male and female and are identified as dual natures.[14] According to Carlo Suarès, who discovered the connection between the ancient Hebrews' concept of spirit and the Egyptians', this same theme is also found in the Chaldean and Hebrew Qabala.[15] (I'll return to Suarès's work later).

The Dramatic Stages of the Soul

The ancient Egyptians based their view of all life on ten dramatic stages of the soul, corresponding to the movement of the sun around the sky. While birth and death were clearly marked, the principle of resurrection was also important and symbolized by the god Osiris, the cycle of the Nile, and the rolling transit of the sun.

Osiris (god) and Isis (goddess) represent the dawning of the world of humanity. In the parable of Isis and Osiris, Osiris was murdered and chopped into fourteen pieces by his envious brother, Set. Osiris and Isis represented natural order and civilization. Set was angry with Osiris for several reasons. Set represented wild, untamed nature and resented civilization. Set was envious of Osiris. Osiris made love with Set's wife, Nepthys. She later gave birth to the god Anubis. The lovemaking between Osiris and Nepthys just refueled

Set's anger. Set knew that Nepthys was pregnant, and he made love with his wife, attempting to skewer the fetus with his penis as he pounded into her. Hence Anubis was born with a jackal's head.

After his death and resurrection, Osiris became a god neither of the earth nor of heaven but of the neterworld, the nebulous world in between. Osiris is the principle of regeneration, while his brother, Set, sometimes pictured as the ultimate materialist, Satan, is the principle of destruction. Horus, the god of rebirth, embodied masculine warrior energy, while Hathor, the goddess of love, embodied the feminine beauty of nature. She is the jubilant celebration of life with feasts, song, love, and dance. Isis (wife of Osiris) serves as the nurturing aspect of wife and mother and the emblem of magical wisdom. Nepthys, her sister, represents sorrow and also intuition. Geb, the father of Osiris and Set, is the earth; Nut, the mother, is sky. All the future descendants of the world are children of Horus, the son of Osiris and Isis.

With all the characters now defined, the Egyptian concept of "soul" comes into being in terms of a conglomerate of characters all seeking their rightful places in the universe. In Egyptian theology men and women are not limited to their mortal shells and spiritual selves, but are complex and interconnected structures in which their physical bodies, spiritual bodies, and mental and emotional states play off one another. Thus, the idea of a soul is not as simple to present, and it is from this Egyptian soul-story that part of our present confusion arises. Indeed, we each have several souls, if we have one, and each of them wants to go to a different home.

A Brief Look at Your Nine Souls

Early Egyptians believed an individual had no less than nine souls, that perhaps one could look at as one soul composed of nine parts, ranging from the most corporeal to the most abstract.[16] Each *soullet* was bestowed by the seven Hathors, the fairy godmothers present at every child's birth. The souls resided in different parts of the body. First was the primordial life spirit, *aakhu*, residing in the blood. Second was *âb*, residing in the heart. Third, the soul *ba* was the ghost that appeared after death and flew in and out of the tomb, often in the form of a bird. Fourth was *ka*, the image one sees when one looks in a mirror. Fifth was *khaibut*, the shadow of the body also thought to be a soul. Sixth was *khat*, the material living body supposedly resurrected as flesh after death. Seventh was *sekhem* representing the vital force in a human being, perhaps resembling the Chinese concept of ch'i.[17] Eighth was *ren*, the secret or soul-name. And ninth was the *sahu*, representing the unification of spiritual and mental powers. Having defined these soullets very briefly, let me go into more detail and give you some examples of how these concepts might be viewed today.

The first soul, the primordial life spirit, aakhu, residing in the blood, was sometimes called *khu*, which means the first soul or the divine intelligence—that which is radiant or shining. Because it resides in the blood, it has a connection with the ba, or third soul, residing in the heart. Khu represents, more or less, the inspirational aspect of life—the message of the gods. As the heart-soul, it appears to have been regarded as an ethereal being—immortal and able to feel emotions. It also lived inside the sahu—the ninth soul or spiritual body. When you feel inspired, your khu is conscious in your body. Basically every church in America and abroad attempts to awaken your khu on Sunday mornings. As you well know, sometimes we find the khu taking a khat-nap (pun intended).

The åb, or the second soul, is the heart, the seat of wisdom, knowledge, and understanding. It links the physical body with the spiritual body. When you are acting from your heart, your åb is conscious. Åb and ba are, of course, related, the former being the container of the latter. This relationship is more intimate than one might suspect. In some sense they are the same entity. The clue to this comes from the fact that they are spelled with the same letters, one the reverse of the other. As we will see later, the name of the soul or spiritual presence is very important; it signifies a deeper meaning and connection to God.

Åb represents what a man may come to know of the world and of himself in silent meditation. Åb is unmistakably the impregnated-with-spiritual-knowledge physical heart. The åb remembers and holds the knowledge of the ba. Just as in the proverbial feeling of seeing a newborn child, the "My heart is filled with love!" feeling tells you your åb is conscious.

Åb is closely associated with khu, thus accounting for the intimate relationship between inspiration and heartfelt goodness, and is held to be the source of both good, when conscious, and evil, when unconscious, found in all animal and human life. The preservation of the heart was very important for the ancient Egyptians. It is also judged on the scales of Maat, the goddess of truth, before the court of Osiris at death. Thus, åb is where guilt or innocence is to be found.

The ba, or the third soul, was defined as the heart-soul of man or that which was noble and sublime. It resides in åb and is connected to ka. It is said that at death the ba flies to Ra—the sun—and resides with Osiris. The ba can take on human form and can be material or not, depending on its will. This could be the cause of spiritual aberrations or ghosts. While åb represents the knowledge accumulated and held in the heart, ba is the eternal and unchanging knowledge. Ba, in ancient terminology, is the closest we have for what we in the West usually mean by soul. The ba can fly like a bird and would be the soul that leaves the body at sleep coming back in whenever it wants to. When one undergoes a near-death experience, the ba is the consciousness involved. Unfortunately, ba is not easy to reach. (I'll have more to say about how the ba communicates with the khat or body in chapter 13 dealing with *soul-talk*.)

The ka, or fourth soul, is the ethereal double, called the *doppelganger* in legends; but this term is often confused with the other spiritual dimensions of humanity. Ka has been mistakenly thought of as one's higher self or astral body, but it is commonly associated with the emotional body of a human being. You have heard people say, when you were feeling out of touch with yourself because you were unusually depressed or angry, that "you just weren't yourself." Indeed you weren't. Your ka was thus aroused and conscious as if a second being had inhabited your body. Football players know their kas quite well. Off the field they are gentle fellows, but in the midst of a game their emotional doubles take over and they become monsters of the gridiron.

As the ka is related to the emotional man and woman, so is the ba related to the universal or spiritual being. The ka is creative and gives rise to movement, while the ba is fixed and eternal. Ka is an abstract personality capable

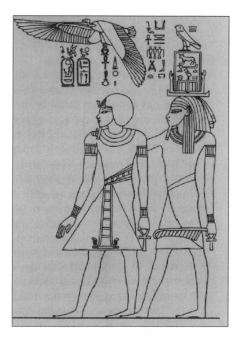

Figure 4.1. The Ka or Double. The ka acted as a guide for the soul.

of assuming form without matter and resembling the person it inhabits. The ka is best illustrated by the movie *Forbidden Planet,* where the good scientist's ka became a monster attempting to destroy the visiting earth spacemen. As in the film, ka can wander at will, but it appears to be somewhat robotic and mechanical, sometimes confusing the real body with a statue of the body, if one has been erected. Imagine a wife married to a Milquetoast character in a Thurber novel who becomes quite surprised when the husband becomes highly charged, and you understand ka.

The khaibut, or fifth soul, is the shadow, and as such it is black or dark in the light. It is literally related to the shadow of a person. Yet it is also associated with the ba, and it is capable of existing outside of the body. The shadow was always something to worry about, in that spirits could capture it, as indicated in the chapter opening epigraph. Today we could think of the shadow in Jungian terms as the opposite side of our best face forward.

In attempting to understand the shadow, or khaibut, one must fully grasp the logical view of the universe taken by mythopoeic humans. There was no sharp distinction between dreams, hallucinations, and ordinary vision. Similarly there was no precise separation between the living and the dead.[18] The holographic notion resurfacing today, *pars pro toto* (the part contains the whole), was well accepted in ancient prephilosophic times. Consequently, a person and his or her shadow were considered connected, and if the shadow was captured, the person could literally die.

People who practice black magic many times use the khaibut as a model of an individual over whom they wish to gain power. As such, the proverbial doll with pins sticking in it represents the khaibut of the individual. When anthropologists first showed photographs of indigenous members of a tribal society to those who had never seen them before, the people whose pictures were snapped became frightened. Their shadows were captured in film and they feared that they would die. In a way they were correct. Western technology has not only captured the shadows of the indigenous world, it has nearly wiped it out.

The physical body or corpse—that which corrupts after death—is called the khat, or sixth soul. It is easy to define and was always associated with the body. Although it never lived very long, it was always animated by ba. Today we might refer to the khat as the "bodymind," to describe the consciousness purely associated with bodily functions such as elimination, autonomic functioning, unconscious breathing, heartbeat, and so forth.

The sekhem, or seventh soul, is the form of a spiritual being, the incorporeal representation of the vital force of a human being. It exists in heaven among the khus and is more or less that power which a man possesses to become incarnate. The sekhem becomes highly aware during the transit phase of existence, between death and rebirth. Its main function would be power and energy to institute reincarnation. Thus, the sekhem functions to maintain vital power in the body.

In addition, a man's name, the ren, or the eighth soul, is powerful and holy. To blot a man's name from history, to forget him, was to destroy him and banish him to oblivion. This emphasizes, I believe, the importance of words in sacred and carnal life. As a man's name goes, so goes his destiny in the afterlife. The importance of naming things is also reflected in the Egyptian Book of the Dead where one finds in Ellis's translation, "that which can be named must exist." Here Osiris reflects on his life and realizes "As I am, I was, and I shall be a thing of matter and heaven."[19] As the soul moves through the neterworld attempting to return to Osiris, the soul, from time to time, is asked to name the various guardians who stand before secret chambers and passageways the soul must enter. Failure to name them or to remember his or her own name forces the soul to return and not gain entrance to the court of Osiris. I take this to mean that if the ren is forgotten, if the soul fails to remember who she or he really is, the soul is banished, most likely to an inferior rebirth as a lower animal life-form.

The ninth soul was the sahu, the spiritual body that sprang from the material body at death. In it, mental and spiritual attributes of the natural body were united and given new powers. The sahu had the same bodily form as the khat but was itself nonmaterial and contained the other soul bodies within. The sahu traveled to higher worlds or, as we would say, was the bodily form of the soul that went to heaven.

A Modern Egyptian Soulful Insight

In writing this book, I realized a complex gap in our understanding existed between the pagan worldview of spirit held by our forebears and our present modern vision of an abstract God and mysterious eternal soul. Subsequently, I had the opportunity to do some research in Egypt and met with several Egyptian scholars, guides, and teachers who helped me to overcome this gap. One of my teachers was Hakim, a guide I first met in Giza. Hakim led our group to as many temples and ancient sites as our feet and minds could handle. According to Hakim, the life of an individual soul corresponds to the stages of the sun as it circles in the heavens.

One of the major misunderstandings concerning the soul can be found in different concepts of time. For the ancients, time was envisioned as circular, whereas for us, time moves along a straight and never-ending line.

Consequently, our forebears saw all things in time behaving quite differently than we would, particularly cyclic behavior. For example, in ancient Egyptian religion one prays according to the position of the sun in the sky.

The sky, represented by the sky goddess Nut, forms a bow across heaven, with her legs on the east side of the Nile and her head and arms in the west. In the morning, corresponding to the first stage of life, Nut gives birth to the sun, symbolizing the incarnation of the soul. Ancient Egyptians believed the birth of a being corresponded to the appearance of the sun and to the scarab or beetle, seemingly arising from its own fecal matter. The Egyptian word for beetle or scarab is *kheper*, and this first stage is also referred to as kheper.

The second stage, *ra*, corresponds to the sun's position midway between noon and sunrise. It can be viewed as the symbol for childhood—the stage marked by joy and happiness. The third stage, *oon*, is marked by the sun at high noon (perhaps we get *noon* from the word *oon*). This signifies the midlife of the individual and symbolizes a growth towards wisdom. The fourth stage is called *aton*, corresponding to the sun's position midway between noon and sunset. It symbolizes even greater wisdom in the life of the individual. Hakim suggested the Hebrew word *adon* meaning "lord" came from this ancient Egyptian word.

Hakim gave me a new twist on an old story, the Biblical Exodus. Interestingly enough, if one looks at the walls of the Temple of Time and Timelessness,[20] called the Ramesseum, one sees a very different impression of the famous Jewish mass departure from Egypt. The story as seen by the Egyptians and told in the bas-relief wall carving of the Ramesseum is that the Jews were not slaves in Egypt. Instead, they were followers of aton and refused to see the soul end its time in the setting of the sun in the west, thus going into the fifth stage called *amon*, from which we get the word *amen*. Aton symbolized the highest stage of intellect, and one might observe the great emphasis placed by Jewish families on their children toward intellectual studies as typifying this. I also heard that some Egyptian sources indicated that Jews in Egypt refused to believe in Amon, however I have been unable to document this.

What were the escaping Jews running from? Hakim explained that they refused to enter necropolis and the neterworld—the Western side of the Nile—and enter into the realm of Amon as their Egyptian brothers and sisters did. And for good reason. Going west meant seeing the end of the sun's trajectory; sunset meant the end of life in the upper world.

Perhaps here in this myth is the division that inevitably separated the Jews from their Egyptian kin. Perhaps the roots of our Western Judeo-Christian culture can be found here. Modern Jews, like most enlightened people, place great value on life. They believe, as most Westerners do, that *this life* is important, perhaps more important than any afterlife we might lead. And *might* is a big word here. Modern Judaism has no answer to the mystery of what happens after death. It remains a mystery to all practitioners of the Jewish faith, with the possible exception of those who follow Qabala (about which I shall have more to say later).

In ancient Egyptian myth, life didn't end with the setting of the sun. It continued into another five stages, ten in all, symbolizing the ensuing life of the soul in the underworld. The soul did not die, it went west where the sun goes when it sets. In Orthodox Judaism and in modern Islam there are just five different prayers every day. These are perhaps related to the five daylight stages of the sun. Nothing else has changed except the names from the early Egyptian religion to the present Judeo-Christian and perhaps Muslim form. What happens after the sunset of the soul?

The Soul's Adventure in the Afterlife

The great mystery was what happened to the soul when it went west. Ancient Egyptians saw the sun sinking in the west as a soul's return to spirit. The cycle of the soul followed the cycle of the sun. According to the Egyptian Book of the Dead, Ani, the scribe, who also depicts the soul inside of each of us, is taken by the hand by Anubis, the jackal-faced god of the underworld, to his final judgment day. His heart is weighed against the feather of truth of Maat, the goddess who represents law and justice and is the wife of Thoth. Maat then entrusts these records to her husband, who carefully writes them down and records the events of the soul's life.

Figure 4.2. The Soul in the Neterworld. The court of Osiris. Ani is taken by the hand to the scale of Maat where his ab is weighed against the feather of truth.

Interestingly enough, Ani in Hebrew means "I am." I believe this is more than a coincidence. Ani is an anagram, and each of its spellings has importance. (Later in the chapter I will explain the sacred significance of such anagrams.) It has often been observed that our ancient forefathers from the land of the Nile liked puns and anagrams.[21] A-n-i is no exception. Change the letters around to a-i-n, and you have the Hebrew word for "there is not." The hidden meaning might be looked at as the Buddhists regard the soul: The soul that says *I am* is not *real* and, ultimately, neither is anything else. (We will come back to this in part 3, chapter 8.)

Thus the trial of Ani is or can be looked at as the trial of humanity—the parts we all have grown to feel and be at home with—our essential selves, if they really exist. If Ani passes this first trial, meaning the soul's heart carries with it no more weight than the truth, Ani is taken by Horus, the falcon-faced son of Isis and Osiris, to the final judgment by Osiris. Depending on the wisdom of several other gods in court, Ani's soul passes through to the neterworld. These other gods reflect aspects of the soul's life, and all have something relevant to say about the continued existence or not of the immortal soul.

Horus is sometimes called Ra or the sun-god. Thus we come again to the heavenly way the Egyptians looked at the life of the soul, following the path of the sun across the sky.

As I mentioned earlier, this solar trajectory, the life of the soul, can thus be looked at in terms of *ten* phases, each phase corresponding to a position of the sun in the sky. The later five phases describe the soul's presence after death, represented by the sun sinking below the horizon and journeying to the other or netherworld. Our ordinary life, represented by birth followed by death, only corresponds to the first five phases: sunrise to sunset. Our life corresponding to death followed by birth is represented by the last five phases: sunset to sunrise, when the sun goes to the "other side of the world." This second half of life/death is mysterious to most of us and was equally so to the ancient Egyptians.

Ancient Egyptians were obsessed with the periodic nature of life and death. You certainly can understand why. Much of their vision was based on the cyclical movement of the rising and falling of the Nile River and their observation of the circular revolution of the celestial sky. The Egyptians believed these periodic movements were strongly connected. Accordingly, everything followed the law of the circle—the movement of the cycle of obvious visible life and secret invisible death.

This ancient—life/death, obvious/secret—vision reappears in our modern worldview. For, as we will see later, when we look at the Dirac equation-model of the electron,[22] the vacuum vibrates and produces repetitious patterns.[23]

The Hidden Soul and the Meaning
of א (Aleph)

Returning to the ancient Chaldeans and their vision based on Qabala, we find a similar view. The soul arose from *nothing-able-to-be-anything*, symbolized by the silent letter א (aleph). Therefore, there was also a hidden aspect of the soul since it came from the great void of the spirit.

Today Qabalists realize that aleph, symbolizing sacred mystery, cannot be completely known. Nevertheless, it is remarkable that the ancient spiritual aleph and the modern physical vacuum are so similar. To gain greater insight into the life and death cyclical myth and its influence on a current quantum physical model of the soul, we need to look at the work of foremost Jewish Biblical scholar, artist, Qabalist, and decipherer of the Hebrew Bible, Carlo Suarès. He was born in Alexandria in 1898 and was so influenced by the Egyptian myth of *time and timelessness* that he used it to uncover a hidden sacred meaning of the Hebrew Bible. As we will see, Suarès's work provides a foundation for a new vision of the soul.

Suarès is a name well known in Alexandria. According to Suarès, he came from a very old Sephardic family that arrived in Spain at the time of the Arab conquest. His family emigrated during the Inquisition to Tuscany and settled in Egypt in the eighteenth century. Many years later, Suarès left Egypt, settled in Paris, and became a French citizen.

I met Suarès in early spring of 1974 while I was a visiting professor at the University of Paris.[24] If there is any truth to the story of the wandering Jew, Suarès typifies it. His illustrious life and travels include an episode that took him from Paris during the German occupation of World War II while his apartment, overlooking the Eiffel Tower, was occupied by the German High Command.

According to Suarès, significant Hebrew words found in the Bible and other sacred writings are also meaningful equations describing the action of sacred forces. Each letter of the *alephbayt* (alphabet) has a specific meaning. When these letters form words, the meaning is enhanced and the action of one letter upon another is spelled out literally. Although it would be too much to go into greater detail about this, I do want to give you a taste of Qabala here. For more, I refer the reader to Suarès's earlier books.[25]

The first letter of the Hebrew alphabet, aleph, represents unthinkable vibration, potential energy, possibilities, able-to-be-ness, timelessness beyond space and time and beyond limits. Any words describing aleph limit it and are therefore not aleph. Aleph is a symbol of the unthinkable.[26] Aleph is also an unspoken letter in Hebrew. It has no sound of its own but takes on sounds according to special signs that are written below it. Aleph is spelled *aleph-lammed-phay (a-l-f)*.

Thus many words that contain aleph also have sacred significance. The word for *word* or *sign* is *aut*, spelled *aleph-vav-tav (a-u-t)*. Sacredly it means aleph or spirit fertilizing its own cosmic resistance, *tav*. The word for *light* is *aur* (much like our word *ore*) spelled *aleph-vav-raysh (a-u-r)*. This means aleph fertilizing the whole universe, *raysh*.[27] The Hebrew word for *earth* is *eretz*, spelled *aleph-raysh-tsadde (a-r-ts)*. Here aleph, acting on the universe, raysh, creates the ultimate feminine structure, tsadde. Thus arises the principle of the earth as a physical embodiment of spirit in a feminine form.

The tenth letter, *yod*, means, in Hebrew, *hand*. During the Olympics in 1972, African-American athletes were asserting themselves as blacks rather than as Americans. I remember one stirring scene as a young black athlete stood on the winners' podium and raised his black gloved fist into the air announcing silently that he was there. The raised fist was a symbol of yod—aleph or spirit in existence. The silent cry, "Brother, I am. Sister, I exist," rang in my consciousness as we witnessed our black brothers and sisters raise themselves from servitude into spirit.

Aleph, projected into material existence as actual space and time as we experience it, is symbolized by yod. Since aleph is timeless and spaceless while yod is its opposite, an antagonism arises between the two. For aleph to be known it must project itself as yod. Thus yod is a sacrifice of aleph. It is aleph, but it has given up its timeless quality. We could call aleph God or the big "I" in the sky and yod the little god in existence. Aleph and yod are thus in eternal conflict with each other. This conflict can be felt in every one of us, as the common suffering we all experience. It is the yearning of the human spirit to return to spirit leaving the material body behind. It is a kind of "death wish," as Freud has pointed out. But it is more than this. It is really a life-death principle. It can be sensed as the "afterlife" wish within us.

Thus the Hebrew letters aleph and yod are in a battle, which is in the Bible itself. One of them, aleph, represents the timeless unmanifested spirit and plays a similar role in the mystical scheme of things as does the Egyptian concept of Atum. The other, yod, represents the projection of aleph into time and is, therefore, the symbol of the soul's existence, its fall into space and time. In the word-equation *Elohim*, meaning God, in the Book of Genesis (spelled *aleph*-lammed-hay-*yod*-mem), we can see aleph in battle with yod, its own projection. The first three letters *aleph*, *lammed*, and *hay* form the word *elloh* or *Allah*, the name of God in Arabic. The fourth letter *yod* comes next followed by *mem*, signifying water and consciousness. Thus Allah directly confronts yod, the hand of existence, with mem consciousness at stake.

As Suarès sees it, the Bible is an instruction book for the attainment of cosmic consciousness. It is written for the soul's instructions, written in code! Deciphering that code is the work of the soul while it is incarcerated in the body.

The Code of the Soul

But why write the soul's instructions in a coded language? One reason can be found in Platonic reasoning. When the soul enters the body it becomes confused. It loses the ability to see clearly, to see with the eyes of cosmic awareness. So which language should be used to instruct the soul while it is in such a confused state? Clearly a language that acts more than it simply describes. For description cannot be complete. It can point to but never replace experience. It can indicate truth, but never be truth. So the early Bible writers would be in a predicament. How can what can't be described be conveyed as a teaching? The answer was to base the Bible in an arcane language with hidden meaning that could only be appreciated through work and study. What was the meaning? *That the soul acts in a concealed or secret manner on matter always attempting to influence existence by bringing spiritual order into the chaotic universe of matter and energy.*

As far as the soul is concerned, there can be only one winner—one outcome to the cosmic drama that spirit and existence continue to fight. Suarès says that aleph and yod engage in the "war with time" we all fight in our everyday lives. This battle has an achievement. It is *qof,* the nineteenth letter of the Hebrew alphabet. In qof the battle is recognized. Consciousness sees itself from its cosmic pinnacle. It plays the game with time but always recognizes itself as spirit. The war thus continues, but the battles are over. Qof is this cosmic principle. It is the symbol of the quantum and the principle of quantum physics—spirit comes into matter and potential reality becomes existence. Qof is the recognition that in order *to be* we must "put up" with matter. We must suffer existence as itself. We must temporarily forget our cosmic selves and in our remembrance of this we return to our sacred heritage. We are released from bondage even though we live in human bodies.

Qof can be taken as the symbol of the quantum—the frontier between aleph and yod. In quantum physics we have a similar mystery-and-the-ultimate-realization-of-the-aleph question: How is it that anything is? Accordingly, matter "pops" into existence as a result of registering in the mind. All that quantum physics can predict with absolute certainty is the probability that anything *is.* Experience tells us that "isness" is business. That things are as they seem to be. The quantum wave function, or qwiff, must "pop" for there to be anything. This popping is the function of qof.[28]

Qof first appears in the Bible in the word *qof-raysh-aleph (q-r-a)* or *qurah,* which means "to label or call a thing by a name." One could see this insight is vital; it is the forming of word symbols. This is the work of the soul. Without naming things, nothing can come into being. This is true, according to the Bible, and, as we shall see next, is also vital to the creation of the universe.

What's It All About, Aleph?

The key insight here, finally, is that the soul arises in the vacuum of space and is continually "fighting" with its own image in matter. The ancient vision of the soul appears to be that it is nonmaterial yet capable of influencing life. The abode of the soul appears to be, from an ancient viewpoint, the unknown and unknowable א, aleph, which projects into existence everything that exists. Aleph, the home of the spirit, has its modern counterpart in the concept of the vacuum. The stories of ancient Qabala and modern physics are similar. Something comes from nothing. From aleph springs the universe as we know it.

This concept is reflected in the Hebrew language. Aleph is the first letter of some important words which signify a bursting forth: these include *love (aleph-hay-bayt-hay)*, *God (Aleph-lammed-hay-yod-Mem)*, *light (aleph-vav-raysh)*, *word or sign (aleph-vav-tav)*, and *earth (aleph-raysh-tsadde)*. For the Qabalists, this means these words are more than words or descriptions, they are energies. In modern physics language, they are symbols of consciousness and energy, equations, ready to manifest, ready to turn chaos into order, and, paradoxically, order into chaos. They are the tools of consciousness.

The battle of time and timelessness, the fight of the order of the eternal with the chaos of the temporal, never ceases. Only in quantum physics can we see this: in the ultimate behavior of the electron moving near light-speed and in subatomic matter. The soul is beginning to show its face from the murky emptiness of space. The soul began to do this at the very beginning of space and time: the big bang. The soul is dancing and spinning the whole universe and is responsible for the creation of not only matter, but also her shadow which is found in the negative energy sea of possibilities called the vacuum of space. (We shall return to this negative energy sea of possibilities in chapter 6.)

If our souls were present at the beginning of time, surely, being immortal, they will be there at the end of time. The beginning and ending of time are quite far apart, however. To see the souls' roles in life to come, it is necessary to speculate about their roles in the future world of science—the world of prediction based on fact. To do this, we need to gain a sweeping picture of the universe from the beginning of time until its very end as predicted by physics. We need to make some appraisal of what the future might hold for us.

In the next chapter we look at this question and how one modern scientist attempts to deal with the soul in a contracting universe populated by computer-generated, virtually-real human beings. Their souls are computer programs resurrected at the end of time.

CHAPTER 5

Resurrection Physics:
A Lesson from the Land
of Joking Smoking Skulls

For the soul is the beginning of all things.
It is the soul that lends all things movement.[1]

—Plotinus

Everything is determined, the beginning as well as the end, by
forces over which we have no control. It is determined for the
insect, as well as for the star. Human beings, vegetables, or
cosmic dust, we all dance to a mysterious tune, intoned in the
distance by an invisible piper.

—Albert Einstein

IN THE BEGINNING OF THIS book, I mentioned the difficulty we have in reconciling science and spirituality. I discussed the growing isolation in our present Western culture brought on by the convenience of mechanical devices, most recently the computer. Perhaps having more machines to accomplish our tasks has not freed us from work; it has filled our time with even more to do and given us less time to contemplate and get in touch with our spirit. It may appear science, coming to our aid by bringing more and more sophisticated machines into our lives, is reducing us to soulless and perhaps mindless automatons, while our machines grow more intelligent each day.

Are we really losing our souls, replacing them with machines? If so, is this really a problem? Perhaps science is leading to a future where our souls indeed *inhabit* the sophisticated mechanical devices of our own invention. As we attempt to scientifically define and prove the soul's existence, we need to consider this possibility next. Perhaps the soul silently speaks to us from the vacuum of space, encouraging us to build these devices in order that we hear her spoken voice, especially as humanity nears the end of its time—an epoch that few of us think about.

If it is possible to build machines with souls (there are none on the market today, I assure you), the construction will take some time. Since it appears our universe will come to an end, certainly human life on our planet will, we must consider the soul's future when the earth has long passed. If we use the past growth in science as our guide in the search for the soul's existence, the future looks quite different from what we might imagine. In this chapter we look at the soul's existence by leaping into the far distant future of our universe: the end of time. We find then our souls inhabiting intelligent, sophisticated computing machines and not a human being anywhere.

This may seem outrageous. How could we ever come to this? The answer is apparent. Could you function today without machines in your life? Hardly a day goes by without each of us making use of many machines. Certainly we all have become dependent on our mechanical devices. Machines come to our aid in new ways every day. Today prosthetic devices enable legless people to actually *run* on resilient artificial legs, unthinkable just ten years ago. Even artificial eyes are becoming possible.

With computer technology's presence in our lives increasing each day, we are learning to interact with computers as if they were intelligent beings. For example, computers can actually respond to the spoken voice and type out the words spoken in any language! For some individuals, notably the disease-stricken physicist Stephen Hawking, the computer has become the only spoken voice they will ever have.

Can computers become anything more than machines? Can they learn to think and respond like human beings? How far can we go in making them intelligent? What happens to us if we can indeed make them in our own images? Would they have souls? Nonsense? Maybe not. I pointed out in chapter 3 through the ancient allegory of Plato's cave—brought forward to our TV room—how many of us are seemingly capable of projecting ourselves into the characters we see on the video screen. Consider for the moment what our future might be if we really could project ourselves into such images. With virtual-reality devices (where the viewer, wearing television goggles virtually participates in a three-dimensional scene by moving, flying, talking to other three-dimensional characters, etc.) and greater reliance on computers and prosthetics, it won't be long before our brains become "hard-wired" to computers.

In other words, we all could become nothing but billions of billions of bits of one gigantic pool of information, structuring itself, rearranging itself, and reconstructing itself into multitudes of patterns: virtual-reality human beings.

Breathtaking, perhaps, but where would the soul fit in? Can a computer have a soul? One might suppose any view of where a computer-resident soul could ultimately be put would be into one of these two camps: Camp A for Aristotle or Camp P for Plato. Remember Aristotle saw the soul as a subtle physical substance, while Plato saw it as a nonphysical idea-like entity. Camp A holds to the material soul, very much an energy and information-based understanding arising from material processes as naturally as sharpness arises from a knife's edge, whereas Camp P holds to a non-material soul, a sort of ghost in the machine, capable of directing whatever body or machine it inhabits into action but also capable of sensing outside a body, without distortion. The P-camp followers, eager for the freedom of undistorted vision, seeking liberty as the slaves of Plato's cave sought release from the chains that bound them, welcome death and eternal peace, meaning escape from materiality. The A-camp followers, rejecting the idealist position of the eternal soul and instead arguing for the soul's disappearance at the moment of death or destruction, are not so eager for the grave or the junkyard. They see only annihilation ahead.

Today, several physicists and technology buffs deal with the existence of the soul[2] and it is not easy to see which camp they fall into. Probably the most outrageous proposal for scientific soul-evidence, one that seems to have feet in both camps or perhaps in neither, comes from a physicist living in the neighborhood of New Orleans—the Mardi Gras city of voodoo ways, a town that, as I described in chapter 1, laughs at death with joking smokers puffing through skull masks on Halloween night as they stroll along Bourbon Street.

Physicist Frank Tipler, at the time of this writing, Professor of Mathematical Physics at Tulane University in New Orleans, in his recent book *The Physics of Immortality*, makes a daring scientifically based proposal, bound to shock some, offend many, and delight, I suppose, the few.[3] Tipler offers no excuses for his bold proposals concerning the soul, God, death, and the Christian form of resurrection. For this he has my admiration. It is difficult to rock the establishment of concrete minds these days—in fact it is in any day, as Galileo, Kepler, Bruno and many others would testify if we could resurrect them. In brief, Tipler says we shall all know life again at the final moment of the universe, and then we shall experience life eternal. That is, we shall be reborn and never know death again. The catch is that we all shall be computer simulations contained in microcomputers, interlinked by light-speed signals traversing the whole universe and spending an eternity of time in just the last few billionths of a billionth of a nanosecond (a billionth of a second) of the universe's existence, just before it all comes to a crunching halt.

That is a tall order coming from a physicist, and what he means by all of these claims might not be as you would hope or expect. Then again, who am I to predict what you would hope for? But, Tipler is no mere speculator in the game of the physics of God and eternity. He makes his predictions based on imaginative but purely physical reasoning—mind you, not the kind of physical reasoning you would use to erect a house or biomechanically compute how to mend a knee thrown out of joint, rather the reasoning of a mind thoroughly engrossed and trained in the intricacies of modern general relativity. This is the relativity Einstein was working on until the day of his demise. Tipler's views are also based somewhat on quantum mechanics and to a much larger extent on computer technology.

I realize that some of you might wonder what this has to do with Christianity. After all, Tipler's viewpoint is scientific, isn't it? So why bring in anything dealing with Christian theology? It turns out that Tipler, in his rendering of nano-technological survival kits, has carefully described the essential elements of Christian theology typically considered in Sunday school and by priests from every Christian pulpit. These include resurrection of the dead to eternal life, what happens after the resurrection, heaven, hell, and purgatory.

There has been nothing like this since Thomas Aquinas based his argument for the existence of God on the physics of Aristotle (see chapter 2). In order to grasp Tipler's sensational vision—one that takes us from the big bang startup of time and space to the "big crunch" signaling the end of everything—we need to understand a few basic concepts.

Tipler's Life, the Universe, and Everything

If present models of the universe are accurate, Tipler's included, the universe will go on expanding rapidly for about one billion, billion years. Then, after a few billion, billion years of slow expansion, catching its breath, so to speak, it will, around ten billion, billion years from now, begin to contract! One might wonder, what about us?

We won't be here then. The earth will long be gone, burned to a crisp by a sun changed into a giant red star filling the space beyond the earth's orbit. But imagine a machine, built by generations yet to come, surviving the burnout of Sol. This machine somehow learns and contains, like an advanced version of our space probe Voyager traveling to and beyond the end of our solar system, all of the relevant information available in the universe. It records everything that has ever happened, every life that has ever existed. Imagine a future when all of our intelligence, feelings, hopes, pains, sorrows, elation, deaths and rebirths, all future generations, all past generations are recorded in the greatest detail possible in a vast number of the most intelligent computing machines yet to be built.

You may scoff at this. But how do you know right now that you are *you* and not a machine exchanging information with itself in some highly complex wetware framework using carbon-based units based on DNA strands? How do you know that you are anything else? Assume that a simulation of your life has been constructed. It is so accurate that anyone investigating it cannot tell it is a simulation, not even you. In other words, no matter how anyone, including yourself, probes, asks, teases, or whatever the device acting just like you, can this prying ever determine that the device is not you? If not, then *it is you.* I'll get back to this, but meanwhile let's carry Tipler's model a bit farther.

Tipler's "Life"

Tipler's vision is truly universal. He is concerned with what will happen to life in the universe not only in the present epoch, but in the extreme future—the end of time. Long before the sun burns out seven billion years from now, humans will have their work cut out. This will mean building and sending tiny, intelligent, communicating mechanical devices into space. He expects that these devices will, in turn, manufacture other devices, like themselves or more advanced, as they fly ever outward in all directions into the expanding universe. To do this, they will need to talk to one another, exchange novel information as required of Tipler's "life."

Tipler defines "life" as *the exchange of novel information developing through time.* This means that if two objects can from time to time exchange new information with each other, they are technically alive. If they continually repeat to each other the same-old, same-old—get caught in a loop, so to speak—then the actions they constitute may not be living, they may be just long-winded recordings talking to recorders that in turn play back the recorded information after a while. Of course they could be alive but leading very boring lives.

This exchange, however, becomes problematic as Tipler's communicating devices move farther apart. Assuming that the devices exchange information by using light signals, the devices must continually wait for each other's answers for longer and longer times as they rocket away from each other. For a truly meaningful exchange to occur, the two objects must be technically within a reasonable range of each other, exchanging novel information. That means that they must be able to talk to each other, have interesting things to say, and enough time to say them, no matter how far apart they are. However, remember the upper speed limit for signal exchange is light-speed and that means, as these objects move out into space and farther from each other, perhaps even thousands of light-years[4] apart, we need to look at how these light signals will behave in an expanding and then contracting universe especially on extremely long time scales.

The Expanding and Contracting Universe

Our present thinking about the universe envisions it beginning at a single point and undergoing a rapid expansion, then after reaching a maximum volume, contracting once again to a single point.[5] In dealing with models of the expanding and contracting universe, we need to grasp a difficult notion: Space itself is expanding and then contracting.[6] Naturally one asks, how can space expand or contract? What does it expand or contract relative to? Atoms themselves do not change their sizes. Yardsticks and miles don't get longer or shorter. Even the distance between the sun and the earth does not increase. So what can be meant by the notion of space expansion?

If one makes a model of the expanding universe as a balloon filling with air, one gets a picture of what this means. If you stick pennies on the surface of that balloon, as the balloon expands the pennies grow farther apart, but the pennies themselves remain the same size.[7] The surface of the balloon contains the whole universe and each penny represents a galaxy. Thus the galaxies grow farther apart while each remains the same size.

If you imagine yourself standing on the Earth while the universe is expanding, an interesting question arises. Suppose I send a light signal into space in the hope that it will reach another galaxy, will it ever get there? During the expansion phase of the universe, the light signal must travel into a space that is ever-stretching ahead of it. Imagine, using our balloon-universe model, that this signal travels outward in an expanding circle, like the circular wave produced in a still pond by a dropped pebble.[8] The universe is, to begin with, larger than the light signal circle. The circumference of this light-circle is restricted to the surface of the spherically expanding universe as if it were a circle of latitude drawn on an expanding balloon. Hence, the circumference of the light-circle defines a boundary separating the regions of the universe enclosed by the light-circle and the regions remaining unilluminated by it. Physicists call the circumference of this light-circle the *light horizon*.

Picture this as a race. The sphere is expanding while a circle of latitude is moving across it like a circular wave on the surface of a pond. This signal, although leaving the earth during our present epoch (15 billion years after the big bang), is, on the time scale we are dealing with (billions of billions of years), quite close to the time of the big bang. It turns out, according to the model, that even though the universe expands, the light horizon will exactly cover the surface of the balloon-universe model at precisely the time the balloon-universe reaches its maximum expansion. To put it another way, the light will have filled the whole universe.

The Big Bang, Big Crunch Model of the Universe

Tipler, basing his ideas on Einstein's equations, proposed that the universe will proceed from the big bang, expand for a while, and then contract to the big crunch—a finale he calls the *omega point* after Teilhard de Chardin, who originally came up with the idea of a final time. Tipler's model satisfies Albert Einstein's field equations. Einstein, you'll remember, attempted to find the meaning of physical life in terms of space, time, matter, and gravity. Throw in some electromagnetic field energy as well, and you have Einstein's dream: a set of mathematical equations that make predictions about the shape of the universe when things began and will end.

Of course the universe can expand and then collapse in different ways. In some of these ways objects can get out of touch with each other, meaning they become too far apart for any light-signaling to reach each other. This becomes a real problem near the end of time when the universe undergoes collapse; if it collapses to a series of separated volumes, communication between the different regions around the collapse points will not be possible. If, on the other hand, the collapse continues to a single point, the omega, communication can continue until the very end.

Tipler's model universe, which he hopes is like our universe, expands and contracts in such a way that failure to communicate will never occur. But as nice and neighborly as this may seem, Tipler's universe will still, unfortunately, come to a devastating end at the omega point. What actually happens is that the universe uniformly expands for a while, reaches a maximum radius, and then uniformly contracts to a big crunch with nary a wrinkle in the fabric of space (meaning no pockets of space form or get pinched off during contraction). But—and this is the big but—if all goes well, we won't know that the universe is coming to a dead end, even though it is doing so, because we will be too engrossed, much too busy to ever notice.

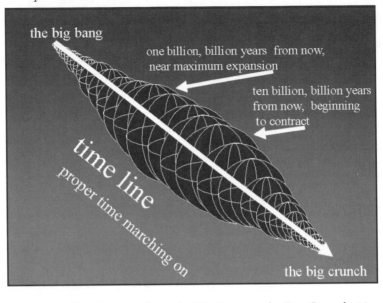

Figure 5.1. The Universe from the Big Bang to the Big Crunch. The omega point occurs at the end of the big crunch.

What will we all be doing? We will all be contained as computer programs in nanotechnologically designed micro-micro devices hardly a few pounds in weight. Our souls will be ghosts in machines, computer analogies of ourselves, and we won't even know it. That is, we won't even remember what we once were when these devices suddenly resurrect us as fleeting bits of data. How can this be? How could *we* ever be reconstructed as fleeting bits of data and not flesh and blood?

Tipler answers that although we as flesh and blood creatures assuredly won't be there, our information will. His answer is based on the assumption that the universe is deterministic. In other words, even though the universe may be chaotic, nothing is ever lost. All of the information, every single bit of it, may be scattered from here to the end of the universe, nevertheless, it is still there. By the time the universe comes to an end—the big crunch finale— the universe will have grown small enough for intelligent machines to fetch and record all of this information.

How? Virtual reality has a lot to do with the answer. Supposedly, the life you now live, those precious moments of pleasure and pain, the sense of self, the pressure of the road on your tired feet, the creak in those weary bones, the elation of a good laugh, the joy of holding a loved one in your arms, all this and more, are just data bits moving around in very complex patterns inside sophisticated "wetware." There is no reason to suppose all of this information flying around is incapable of being reproduced by dry silicon units instead of wet carbon-based units. No reason a virtual-reality simulation of life cannot be exactly the same as "real" life. No reason to suppose a computer, complex enough, could not emulate, synthesize, and "create" "you" based on incoming information.

There is no computer like this today, and there won't be unless intelligent life-forms start building them. Since we don't know about other life-forms in the universe, we humans better start constructing the prototypes now, envisioning a time when the future generations of these early models become so complex that they, in turn, will be able to build even more complex future computers. These "final days" computers must be complex enough to calculate life, the universe, and everything. Of course, to do this the computers must be composed of atomic or molecular-sized switches and the like, and they will need to be able to talk to each other over vast distances by light-signals. Even more, they will need to be able to build other devices like themselves even more complex than they are. In other words, these communicating devices will form a gigantic, weblike, self-replicating computer as big as the universe itself and growing as the universe itself expands and, more important, communicating more quickly as the universe contracts.

In Tipler's vision, we must create nanotechnology with all of the mind power we can muster and ship it into space. Eventually, these computers will build others more advanced than themselves. These higher generations of silicon-dioxide intelligent life-forms, surviving our present carbon-based life-forms by escaping the confines of a doomed Earth, must continue to build more and more, tinier and tinier machines as they themselves are hurling through space. They must grow smarter and smarter as they continue their mission: to fill the universe with intelligence.

Like spores released from an exploding mushroom, we can imagine a spherical wave of these intelligent computers traveling outward from the earth in microspaceships, ever expanding as they propagate their kind into the universe. Tipler calls this the *"life" horizon*. We can imagine this "life"-wave traveling outward at a significant fraction of lightspeed. Tipler speculates that in due time the "life" horizon will reach 90 percent of lightspeed. Eventually, soon after the universe begins to contract, the "life"-wave will catch up with the light-wave started much before it.

The really interesting "stuff" starts happening as the universe contracts. According to Tipler, not until the very end will these microcomputers be close enough, fast enough, small enough, and numerous enough to reproduce all the databases ever created. Tipler calls this last bit of omega, the resurrection of eternal life.

How Heaven and Hell Arise Naturally
at Omega in Tipler's Model

As the universe goes from birth to death, heaven and hell arise as naturally as apple pie at omega in a complex of highly speeded computerized nanotechnology that reproduces, at the final instant of time, a virtual reality of everything that ever was. All we need is for life to be simulated by programs, a kind of virtual reality on the scale of microns and nanoseconds[9] rather than on the human scale of meters and seconds. This means that these computer simulations will be processing information extremely quickly. In fact, as the universe approaches omega, when everything that once was far apart gets extremely close, things will be speeded up even more into a kind of complex, gigantic but ever-decreasing-in-size universal, computer-generated, virtual, orgiastic frenzy.

This is Tipler's vision of heaven. It's in the last few billionths of a billionth of a billionth of a second heading toward omega that the fun really happens, and we all get resurrected in a virtual-reality simulation carrying out all of the lifetimes of humanity and perhaps even enjoying other previous life-forms. We also will live through and repeat all of the hells and wars we have created. There is nothing, in principle, left out.

Time Gets Kind of Funny at the End of Time

There is plenty of time for the resurrection of every soul that has ever lived because at the end of time, versions of everything that ever happened during the universe's long life are scaled down and speeded up. But with everything going so fast, how much fun would it be to eat a simulated slice of mom's best apple pie in a few billionths of a nanosecond? Well, time gets kind of funny in the light of General Relativity theory when things near the end of time. It's all relative, you see. It seems that time really stretches. A few seconds on a hot stove is nothing like a few seconds in your lover's arms. We only know life on the scale of a human lifetime, but if we could scale the processes of life down to the billionth-of-a-second range so that one nanosecond of simulated experience was like one second of actual experience, the simulations would never know any difference! So one minute of proper time near omega will actually appear to be about twenty thousand years of experienced time for those that are exchanging information.

Some of you might wonder at that. Life in just a few nanoseconds. Well, remember, everything is relative. And if things speed up faster and faster as omega approaches, no one experiencing anything will notice because those final nanoseconds will appear to last an eternity.[10] We can almost see how this might occur. Right now we speak to each other over telephone lines. It takes about a half-hour to have a reasonably intelligent conversation about life, the universe, and everything. But, with high-speed computers talking to one another, the time to have this half-hour conversation is considerably speeded up to just a few, if that long, nanoseconds. In other words, two computers could say to each other, if they were as intelligent as you and your friend, in perhaps one nanosecond, the same things that you and your friend say to each other in thirty minutes.

To grasp the significance of this, consider that one billion seconds is just under thirty-two years. So one nanosecond is to one second as one second is to thirty-two years. That means that in just over three seconds two intelligent nanotech life-forms could carry on more than one century's worth of human conversation.

From the Big Bang Until the End of Time

Tipler's model shows the universe in various stages of development, from 10 million, billion years from now, when the universe is 3,000 times larger than at present; to 100 million, billion years from now, when the universe is still expanding and is about 30,000 times larger than the universe today; to 1 billion, billion years from now, when the universe is still expanding, has nearly reached maximum size, and is about 300,000 times larger than the universe today; to 10 billion, billion years from now when the universe contracts, and is about 250,000 times the size of our present universe.

If we return to our balloon-universe model, we can once more see an expanding and contracting universe and Tipler's idea of the two circular horizons: "life" and light. During the expansion period, as time passes, the balloon (universe) expands. The expansion shown in figure 5.2 displays the universe when it is about 3,000 times larger than it is now. Although 10 million, billion years have passed, the universe has not reached its maximum. We can still blow more air into the balloon.

"Life," meaning computerized novel intelligence, moving at 90 percent light-speed, has not filled the universe. Tipler expects that these superfast but tiny computing machines spreading outward on the "life" horizon will be able to simulate all the souls that have appeared from the big bang to the omega by the time the universe reaches the end of time.

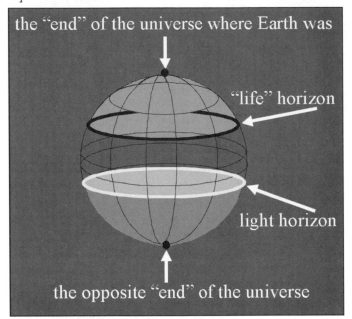

the "end" of the universe where Earth was

"life" horizon

light horizon

the opposite "end" of the universe

Figure 5.2. Tipler's Universe 10 Million, Billion Years from Now. The universe, shown as the surface of a sphere, is 3,000 times larger than present. The earth is shown at the north pole of the sphere. The dark cap covering the southern hemisphere is beyond the light horizon, just south of the equator. "Life" has expanded to about the 45th degree northern latitude and is shown as a shaded cap with a black circle around the Northern Hemisphere, while light which left the earth in the twenty-first century has reached the latitude just south of the equator (marked by the light horizon arrow)—the known "edge" of the universe.

Remember, the light horizon I described above is the location of an expanding "circle" of light that began as a point just about the time the big bang began. In figure 5.2, that light horizon is represented as a circle of latitude on the spherical balloon-universe. At this time the horizon has reached just south of the equator of the universe—the circle of latitude of greatest radius and therefore the "radius" of the universe.

Since 10 million, billion years into the future the universe is about 3,000 times larger than now, it will stretch across roughly 45 thousand billion light-years. Thus, from the Earth, light will take about 45,000 billion years to reach the horizon. We, hopefully building nanotechnology like mad and having its progeny shooting it out there at 90 percent light-speed, will have expanded our "circle" of knowledge, the "life" horizon, to about 22,000 billion light-years by that time, but still there will be a long way to go.[11]

By this time, the Earth will have long ago ceased to exist. Seven billion years from now the sun will expand and engulf the Earth, so we, if we are still around, will have long ago left the nest and made our way to another solar system, another star.

In 100 million, billion years from now, shown in figure 5.3, the universe will be really big—30,000 times larger than now—and a long time will have passed since the last time we looked. It's hard to imagine the scale I'm talking about here. Assumedly "we" (meaning our intelligent machines representing us, since unless we left the earth we will have long been gone as flesh and blood entities) will have grown quite a bit smarter than we are now and will have learned how to compute like crazy. Even though the universe is ten times

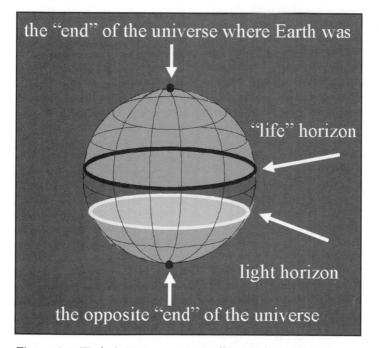

Figure 5.3. Tipler's Universe 100 Million, Billion Years from Now. The universe is still expanding and is about 30,000 times larger than the universe today (although it is shown here as the same size as in the previous figure). "Life" has expanded to about the 45th degree southern latitude, while light from the twenty-first century earth has just about reached the 75th degree latitude near the universe's antipodal "end."

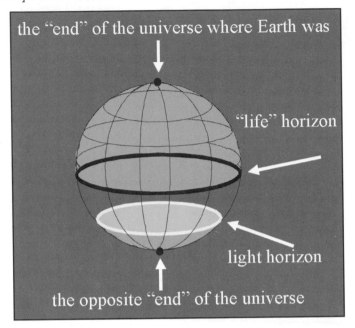

Figure 5.4. Tipler's Universe 1 Billion, Billion Years from Now. The universe is still expanding, has nearly reached maximum size, and is about 300,000 times larger than the universe today. "Life" has almost caught up with light and expanded to about the 60th degree southern latitude, while light from the twenty-first century earth has just about reached the opposite "end" of the universe.

what it was the last time we looked, our "circle" of "life" has nearly reached the light that left the earth so long ago.

In figure 5.4, we see the universe 1 billion, billion years from now at just about maximum size. "Life" has nearly caught up with light. In other words, intelligence is slightly behind light-speed in its race to the ends of the universe.

In figure 5.5, 10 billion, billion years from now, the universal contraction has begun like the contraction starting birth. "Life" and light are essentially the same in that their "circles of influence" have reached the whole universe. In fact the light that left the earth in century twenty-one is now returning to where the earth once was. It will be a long time until omega, nearly another 10 billion, billion years. And during most of that time things will gradually speed up. Only during the very last billionth of a billionth of a nanosecond will things get interesting. Only then will resurrection occur and

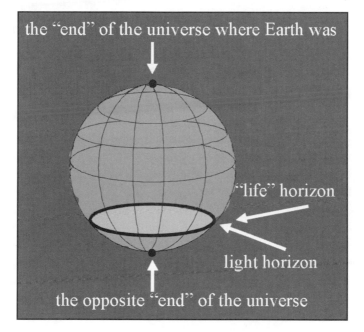

Figure 5.5. Tipler's Universe 10 Billion, Billion Years from Now. The universe is now contracting, and is about 250,000 times the size of our present universe. "Life" has caught up with light, which is now making its way back to the earth (marked by the dark circle of southern latitude), and the whole universe is "alive" with intelligence.

will we enjoy eternal life, even if it lasts only for less than a nano-nano-nanosecond of time.

As bizarre as this might seem to us 20 billion, billion years before then, it appears such a vision is not too far from one envisioned by . . .

A Tibetan Lama in Cyberspace?

In a book[12] recently published by Jeremy Hayward and ecobiologist Francisco Varela, the authors interviewed His Holiness, the Dalai Lama, on topics dealing with science, the human mind, and Buddhism. In one of these conversations His Holiness asked about the physical makeup of modern computers: were they made of metal, plastic, circuits and other apparently inorganic materials? This curiosity had to do with the possibility of a soul's entering a complex computer.

Varela responded that indeed modern computers are constructed of the most contemporary-type materials, to which I would add that the most important ingredient is ordinary sand!! From sand, silicon dioxide, magical computer chips are constructed, chips that carry out millions, if not billions, of instructions per second. Varela then explained that the pattern is most important, not the material from which the computer is fabricated. This pattern is the *program* running the show, telling the central processing units where to put which bits and for how long, and what to do next, and so on.

Indeed the computer today is remarkable and even seems human at times. For example, it will recognize your speech patterns and turn what you say into alphabetic characters, seeming to recognize words and understand them. When asked what he thought about artificial intelligence, the Dalai Lama made an astonishing statement:

> It is very difficult to say that it's not a living being, that it doesn't have cognition, even from the Buddhist point of view. We maintain that there are certain types of births in which a preceding continuum of consciousness is the basis. The consciousness doesn't actually arise from the matter, but a continuum of consciousness might conceivably come into it.

Hayward questioned this. He wanted to know if this meant that if consciousness existed it necessarily had prior continuity with another consciousness—that whenever cognition took place, this cognition was only possible if it had such a connection, a streamlike causal emergence with consciousness starting at the beginning of time?

The Dalai Lama explained:

> There is no possibility for a new cognition, which has no relationship to a previous continuum, to arise at all. I can't totally rule out the possibility that, if all the external conditions and the karmic action were there, a stream of consciousness might actually enter into a computer.

Varela and Hayward appeared to be taken aback by this bizarre statement about the consciousness of a computer. His Holiness continued:

> Yes, that's right. [Dalai Lama laughs.] There is a possibility that a scientist who has very much involved his whole life [with computers], then in the next life . . . [he would be reborn in a computer], same process! [laughter] Then this machine which is half-human and half-machine has been reincarnated.

Although in parts 3 and 4 of the book we will look more closely at the Buddhist concept of mind, matter, soul, reincarnation and the like, I must admit feeling a very strange sensation in my gut when rereading the Dalai Lama's views concerning consciousness. I hadn't realized how close his views were to Professor Tipler's.

The Dalai Lama says that consciousness simply continues based as it was on whatever was happening to the conscious entity before it stopped or, in the case of human beings, died. Thus, even though you die and are dead for a very long time, you will never know it when you resurrect at omega. It's like going in for surgery and being put under general anesthesia. One moment you are awake awaiting the surgeon and the next you are awake and recovering, perhaps even hours later, from the surgical procedure. It's as if nothing happened in between.

In spite of this, I still think that both Tipler and His Holiness are missing something very important. Mainly myself. Now, when you read the word *myself*, apply it to *you*, not to the author! When I write I, think of the pronoun applying to yourself, and read to the end of this chapter with that insight in place.

Where am I in all of this? And where will I be found 20 billion, billion years from now when the omega brings this universe to a close, shrinking it smaller than the period at the bottom of the question mark ending this sentence?

I have a feeling, an intuition, that I have always been here, and that I always will be here. In the next few chapters, we will look at how the ancient vision of I and a modern picture of soul come together.

wickedness is actually bodily disease and not psychological illness? Plato points out that evidence exists to the contrary, as I made clear earlier. But even if wickedness is a bodily disease, using the specificity argument, it would have no power to kill the soul.

Consider guilt in light of this argument. Suppose a person is so wicked that he suffers guilt because of this wickedness. Could guilt, admittedly a soul's evil, cause the body harm? Following Plato's argument, guilt has no power to kill the body even though guilt could cause the body to react, producing its own responses that were death-dealing, such as a stroke, or heart attack. Then guilt also cannot destroy the soul.

Plato emphasized that each soul is inviolable. Therefore, the number of souls is also inviolate; the same number of souls must exist because no soul can die and no soul can be born or created. If the number of souls could change, where would the change come from? Any increase in the number of immortals would, of necessity, come from a vampire-like strike on the mortals. In that case even the mortals would be immortal. Then the soul would be complex, part mortal and part immortal.[16]

Thus Plato's concept of the soul renders it immortal and if not for its constant wear and tear suffered at being enclosed in a human body that does suffer destruction, the soul would be seen in all of its beauty and purity. As far as Plato's soul is concerned, if whatever happens to it doesn't kill it, it must grow stronger.[17] Hence the soul's journey is one that leads to its eventual enlightenment and moral strength. It grows stronger, even though it becomes polluted when it enters the space-time of the body, and recovers when it returns to that space and time that lie outside of space and time.

Where would the idea of purity of the soul outside time and space arise? And why should entering into space-time cause it so much trouble? It turns out that we need to go back even further in time to find the answer. It began with the ancient Egyptians, Chaldeans, and Hebrews. We will look at their ideas in part 2, chapters 4, 5, 6, and 7, where we jump from Plato into the world of modern quantum physics and a prephilosophical world older than Greece as we continue our attempt to define the soul and seek the right questions to answer.

CHAPTER 6

Mind, Soul, and Zero-Point Energy

I am not a human being, I am a human becoming.
—Samuel Avital

Vainly I sought the builder of my house.
Through countless lives
I could not find him
How hard it is to tread life after life!
But now I see you, O builder!
And never again shall you build my house.
I have snapped the rafters,
Split the ridgepole
And beaten out desire.
And now my mind is free.
—Dhammapada, *The Sayings of the Buddha*

ACCORDING TO OUR PRESENT UNDERSTANDING, about 15 billion years ago, give or take 5 billion years, the physical universe was created from nothing. It seems there was a big bang and some time later, perhaps around 20 billion, billion (2×10^{19}) years later, the universe will come to an end in a big crunch.[1] Such a prediction comes straight out of the mouths of physicists who solve Einstein's General Relativity equations.

How about the spiritual universe? Did it have a beginning? Will it come to an end, perhaps as Dr. Frank J. Tipler predicts, with all of us resurrected as computer programs in tiny but linked nanointelligences? Maybe that's what we really are, tiny molecular machines processing data in endless streams in impossibly rapid actions taking only billionths of billionths of a second to complete, just having the illusion that we are human beings living normal lives in space and time suited to our measure. Maybe we all live in virtual reality in some gigantic computer program as big as the universe and designed by God. And then again maybe this is just the illusion of a mind engrossed with computer technology!

Assuming that we are more than *its made from bits*,[2] and that there are both physical and spiritual universes—one that is really "out there" and the other somehow really "in here," inside of each of us—we might wish to inquire how the soul fits into this *big bang* and *big crunch* universe, from a different point of view than that offered in the previous chapter. Consider that Tipler's "soul," being nothing more than a nonmaterial computer program, a pattern in the machine, is based on a Platonic viewpoint. What about Aristotle's concept of the soul as being physical substance? Is the soul a *physical* process, not just a computer program? Or can it be simultaneously real and not physical? Does the universe gush forth soul as it brings forth matter and energy? Does the soul require energy? Or is the soul some form of energy itself?

Something else is at stake here. If it turns out that the soul is physical, a new vision of the universe may appear—nature not only produces matter and energy from seemingly nothing, but also creates soul. But where is the soul? And how could nothing just produce something, anything at all?

No problem for the ancient Chaldeans and Qabalists, as we have seen. According to them, it just happened! There is nothing more that can be said. But why should it just happen? Why should nothing suddenly produce something? Perhaps it's because nothing is really something after all.

All and Nothing at All About Soul

Aristotle once wrote that nature abhors a vacuum. And until very recent times physicists have believed this dictum. However, recent advances in our study of "empty space" have led to startling new discoveries indicating that a vacuum is not nothing. According to quantum theory, a vacuum, consisting of the space between subatomic particles of matter or between large gravitating bodies such as the earth and the sun, is not empty, but consists of vast amounts of positive and negative fluctuating energy. Thus, out of a vacuum can be derived a number of unusual phenomena, including matter, antimatter, energy, and now, as I suggest, even spirit and soul. To understand this, we first need to ask the following: How can energy and matter come from nothing? The answer turns out to be, well, nothing *is* something, after all, and that something, called the vacuum state by physicists, turns out to be very unstable. It tends to boil out matter and energy like bubbles appearing in a raging liquid.

This answer has relevance for the appearance of the soul: The soul, too, comes into existence in the same way.

What Is a Vacuum If It Is Not Nothing?

The concept that a vacuum is not empty seems to be a paradox. If it is not empty, then what fills it? Physicists define a vacuum in very pragmatic terms: It is what you have left after you have removed every bit of matter and energy from an enclosed container. When all energy and matter have been extracted, seemingly magically, there is still something left.

What is left is very interesting because one must revert to quantum physics to deal with it. Physicists call it the *zero-point energy*. Thus, we will look at the vacuum and the zero-point energy to find the soul, that is, if physics has anything at all to say about the soul.

What Is Zero-Point Energy?

Zero-point energy wouldn't have even been thought of if it hadn't been for quantum physics. According to classical physics, things *have* energy. They are able to absorb and emit this energy in the form of electromagnetic radiation. Accordingly, once a source of radiation is turned off, the radiation field, filling any space it happens to be passing through, just disappears. Of course, if there are any reflecting devices in that space, such as mirrors, the radiation can be trapped between the mirrors. If the mirrors are shiny and cold enough the reflection can continue unabated for a long time. But eventually all of the classical radiation will be absorbed.

Not so for a quantum-physical radiation field. Here the uncertainty principle of Werner Heisenberg surfaces and rules the game.[3] Accordingly, even if you extract all energy from a vacuum and suck out every bit of matter, and even cool the vacuum to absolute zero temperature,[4] an enormous amount of energy, called the zero-point energy in reference to that point on the absolute temperature scale, will remain in the vacuum.

Let me give you another example of zero-point energy. If you take a pendulum and set it swinging, it will oscillate for what seems to be a very long time. But eventually, it will lose all forms of visible energy and the amplitude of that swing will decay until the pendulum comes to rest. At least that is what we think happens when we envision the pendulum in terms of classical physics. But in quantum physics, that never quite happens. When the pendulum reaches its quietest moment, it will still be jiggling randomly about its resting position. And even though we cannot observe this motion when, for example, looking at a massive grandfather clock, this jiggling will be present even if the pendulum and its surrounding environment are supercooled to the absolute zero of temperature. Quantum jiggling is always present.

Quantum physics implies a basic fuzziness for all matter and energy. This blur appears as slight fluctuations in the energies or positions of all atomic and subatomic matter. Consequently, there is always a slight probability that any atom will change its energy unpredictably. When an atomic system reaches its lowest energy, one would think that it would come to rest. But this is not so. Instead, the atom or system of atoms continues to jiggle around because of its continual interaction with the zero-point energy of nothing.

This zero-point energy, although it may be impossible to measure directly, is also subject to the laws of quantum physics and it, too, consists of random fluctuations. The amount of energy in those fluctuations is amazing. If one looks into a cube of space, slightly smaller than a die on a Las Vegas crap table (one cubic centimeter), one finds an energy equivalent of 10^{94} grams of matter.[5] That is a lot of mass/energy! Consider that the sun radiates away only the equivalent of 5×10^{12} grams, five million tons of mass, every second, in order to provide you and me with light and heat. That's a lot, but compared to 10^{94} grams it is truly insignificant.[6]

What Is Matter?

With so much mass, why don't we see any of it? Particularly since this mass/energy is fluctuating so wildly, we might expect to have some evidence of its existence. Actually, we do! The vacuum fluctuations of zero-point energy/mass[7] conversion are in the background static on a mistuned radio or on a television screen when a station goes off the air. Even so, it is of hardly any consequence and certainly not as dramatic as the nuclear fusion reactions occurring in the interior of the sun.

Physicists deal with existential questions not so much in terms of defining what something is, but in defining how it is observed to behave. Generally, when we say we are observing matter, we are actually observing whether or not something is changing its energy. We call that something that changes energy matter. However, the division between matter and energy was wiped out many years ago when Einstein proposed, based on some very unusual arguments regarding space and time, that energy and matter were actually the same thing.[8]

Heretofore, people believed matter was "stuff." It consisted of something. As science progressed, the search was for the ultimate stuff of matter. As experimental methods improved, exploration began of the minutest corners of space and the tiniest fractions of time. Matching the search, physicists discovered that matter, seemingly so solid when observed on a scale of inches and seconds, was actually mostly empty space with tiny bits of something called probability clouds filling that space. These clouds are fundamental particles' "tendencies to exist," to somehow pop out of nothing and become *coherent* matter. Matter's ability to *cohere* provides the stability needed to form atoms and molecules.[9]

The search for ultimate matter continues.[10] We now know that if enough added energy is present, empty space itself is capable of spontaneously giving birth to bits of matter and antimatter that will persist for long periods. But even if extra energy is absent, the vacuum will spontaneously create electron-positron[11] (the antimatter form of an electron) pairs and then quickly bring them together, causing them to vanish. Thus, we suspect the vacuum produces matter from the zero-point radiation field, or, in other words, matter is just another temporary state of the fluctuating fields called the zero-point energy.[12]

The Soul: Swimming in a Sea of Negative Energy Electrons

If you stop to think about it at all, you might realize that life on a planet as we live it is really a surprise considering just how empty the universe really is. In fact, the universe is more than 99 percent nothing! And considering that the universe is still expanding at an alarming rate, it's getting to be more nothing than it ever was! So while looking out at it leaves us in awe, when we consider the microworld of subatomic matter, it's even worse. There, nothing exists in spades, so to speak. Probability clouds govern whatever fleetingly exists and even the most refined structure of the atom turns out to be a structure composed of vacuum space.

As far as we know, quantum physics governs the behavior of that vacuous atomic structure consisting of nuclear particles bound tightly together and subatomic electrons moving or not moving, depending on how you wish to think about them, in vast spaces surrounding the nuclei. When one attempts to deal with these electrons, one finds that if they are imagined to be moving, they must be moving really fast, so fast that one needs to consider Einstein's relativity in their peregrinations.

Nice if one can do it, but so far, with the exception of one early successful effort, this has been very difficult to do and still has its problems.[13] That exception is the theory of Paul A. M. Dirac, who was the first to consolidate relativity and quantum physics in dealing with these fleeting electrons.

As Richard Feynman put it:

> Dirac, with his relativistic equation for the electron, was the first to, as he put it, wed quantum mechanics and relativity together. . . . The crucial idea necessary . . . was the existence of antiparticles.[14]

Dirac began to consider the electron when, in 1928, there wasn't much else to consider. Only three particles were known, the electron, the proton, and the photon.[15] (Now there are some 400 or more exotic forms of matter skirting around nuclear corners of space.) Dirac was attempting to build a mathematical structure that would allow the electron to move with near-light-speed, as it had to do when it was close to a nucleus with a large atomic number. Dirac's concern was the mathematical symmetry of the equation describing a single electron, when you put relativity and quantum mechanics together—something he called the equation's "beauty."

This expression, now called Dirac's equation, shows that surprisingly all electrons, following jagged paths through space, move at the speed of light. This "jitterbugging" or zigzagging motion produces the illusion that electrons move slower than light because, instead of traveling on straight lines from any point to any other point, they must fluctuate and, hence, deviate from following the shortest path. This increases their flight time and makes their apparent speed slower.

In fact, an electron could move at the speed of light and go nowhere. Imagine running as fast as you can on a crowded street to catch a shortly departing bus ahead of you. To reach your destination would be simple and timely if there were no obstacles in your way. If you could run in a straight line, your speed would easily enable you to catch the bus before it leaves. But having to change your direction several times because of various obstacles along your way, even having to run back in the direction you came from, you not only miss the bus, you find yourself back at the starting point of your journey.

Why should an electron behave so bizarrely? The answer, it turned out, was the electron was in continual interaction with other "potential" electrons in the vacuum, as if the vacuum was a reflecting mirror with an infinite number of images battering the electron backward and forward. Dirac showed that the vacuum contained an infinite number of electrons with *negative energies*—below the threshold of any measurable perception—and that these negative-energy electrons must exist in order for a single electron to appear in the universe with positive energy. In other words, the vacuum of space holds these electrons as potential, but not manifested, matter. Only when certain energies are added to empty space will one of these unmanifested electrons reveal itself

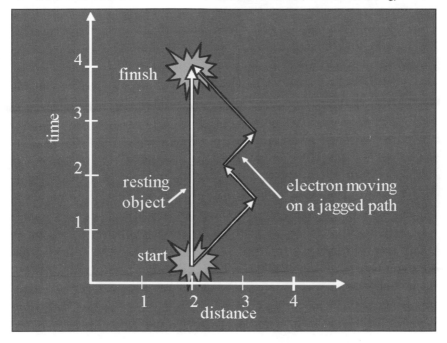

Figure 6.1. A Jitterbugging Electron. The above shows a space-time diagram with time increasing upward. The scale is quite tiny, involving billionths of a billionth of a second, and billionths of a meter. An object at rest traces out a vertical straight line from the starting time to the finishing time, indicating that it has maintained the same position while tracing out a path in time. A moving object traces out a diagonal line, either to the left or to the right. Here we see at one starting point an electron and another object. The electron leaves, moving to the right while the other object remains at rest. After traveling some distance to the right, then some distance to the left, and again, in zigzag light-speed movements, it arrives right back at the same place at a later time where it again encounters the resting object. Although the electron has moved at light-speed, because of its zigzag movement, it has ended up going nowhere. It appears as if it hadn't moved at all. Moving at light-speed and going nowhere, perhaps the soul, like an electron, rocks and rolls inside you.

and leave behind a hole in nothing! In the topsy-turvy world of quantum physics, a hole-in-nothing turns out to be something. This hole has physical properties and appears as a real *positron*—an electron made of antimatter.

One could imagine the vacuum as a sea of potential matter with negative energy and a real electron as an evaporated drop of that ocean with positive energy. Each electron, having positive energy, could fall back into the sea of negative energy as easily as a book tumbles from a shelf to the ground. Electrons lose energy by emitting light, if a lower energy state is available to them. All atomic electrons behave in this manner whenever they are in excited or

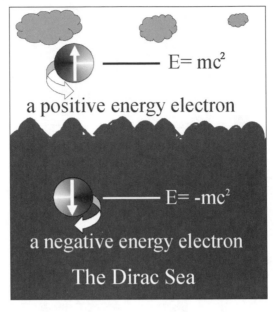

Figure 6.2. The Dirac Sea. A single electron has
popped out, leaving a hole-in-nothing behind.

high energy states. The fall of an atomic electron from an excited state to a
lower energy state produces a characteristic light as seen, for example, in neon
lamps and lasers.

If there are negative energy states in the vacuum, why didn't every manifested electron having positive energy fall back into that sea of nothingness,
giving up its energy as light? In other words, why were there any electrons at
all? For that matter, if Dirac's equation applied to other particles too, why
were there any particles in the universe? Shouldn't they all be sucked into the
sea, giving up their energies as well! If this happened the universe would not
contain any matter at all, just radiation.

Here Dirac was stumped until he met with physicist Wolfgang Pauli.

Pauli pointed out a new principle of physics, later called the Pauli Exclusion Principle, which explained a peculiar property of an electron: It,
electron A, can't fall into a lower energy state if that state already has another electron (electron B) in it unless B is spinning in the opposite direction of A. All electrons exclude each other in this manner. No two of them
ever enter the same state of energy and maintain the same spin-direction.
Like having only so many books on a shelf and no more, such exclusion
took up space so the vacuum of space, the sea, was filled with oppositely

Figure 6.3. The Dirac Sea Is Filled. With the sea stuffed with oppositely spinning negative energy electrons, all obeying Pauli's Exclusion Principle, no more positive energy electrons can disappear into the vacuum.

spinning potential or *virtual* electrons—all that the vacuum could bear. On occasion, when one popped out, changing from a virtual electron to a real one because enough energy was made available to it, it would leave behind a hole like a ghost that would migrate with the opposite character. In other words, the hole in the Dirac Sea would behave as a real particle of matter with the opposite electrical charge of the electron moving through the vacuum of space.

This particle—the hole in the Dirac Sea—was called a *positron*. If an electron moved clockwise in a circle about a magnetic field, the positron moved counterclockwise. If the electron interacted with the positron, it appeared as if the electron was interacting with its mirror-image self, having a positive mass and positive charge opposite to the electron's negative electrical charge. If the electron happened to fall back into the sea and fill up the bubble-space, the electron and the bubble would instantly vanish, giving up an energy equal to twice the mass-energy of the electron. This was called electron-positron pair-annihilation. Hence this notion of antimatter, particles capable of annihilating matter, was born.[16]

Even though the Dirac Sea model of the vacuum shows that the vacuum is empty, it has an effect on everything contained in it. At first glance the vacuum is just space. But when you get down to a finer scale, this vacuum is boiling into spontaneous eruptions of electrons and holes that, vanishing as quickly as they appear, return from whence they came. This constant percolation, the virtual interaction of electrons with their antimatter reflections deep within the vacuum of space, gives rise to the common static you hear on your radio and is the ultimate source of everything that is.

The Electron and I

But along with this rather romantic notion, something else was brewing. A new vision of mind in the universe was appearing. Let me explain. In the beginning, according to all that we know about such things, there was nothing, nothing at all. But then, something rather miraculous happened. Matter, antimatter, and energy suddenly spewed into existence and remained! It was a kind of fluke as if the universe-of-nothing had nothing better to do and so created something. According to our best calculations, if we add all the energy in the present universe, including the energy contained in matter and antimatter, and including the attractive kind of energy in gravity, we come up with a big zero. It all adds to naught. Funny, this nothing business: even when you think you have something, you don't.

The empty vacuum out of which everything manifested seemed to be able to create energy as long as it returned it to balance the books. This balancing act could be viewed in terms of the ancient Chinese concept that the vacuum energy was ch'i—a mysterious energy which is also known as the void of Buddhism out of which everything arises, including our thoughts and feelings. Consequently, like the bubbling Dirac Sea, one could view ch'i as energy in continual interaction with all things, including the human mind. Indeed some view ch'i as mind itself.[17]

One scientist taking this point of view seriously is Dr. Shiuji Inomata, a senior scientist at Ministry of International Trade and Industry Electrotechnical Laboratory in Japan. He sees our present soulless scientific venture as dangerous because we fail to study ch'i with serious scientific attention. Scientists like Inomata believe in the principle of a universal mind—a vast holographically structured mega-computer (God). They see the vacuum as linked to our thoughts and feelings as much as it is linked to the appearance of matter and energy.[18]

I suggested a similar possibility in my earlier book, *Star Wave*.[19] Consider another like view taken by French physicist Jean Charon:

If I suppose that the electrons constituting my body not only carry my "mind-spirit" but, in fact, constitute it, then there would be no difficulty in acknowledging that my "I" or mind-spirit and my electrons are in communication. There is identification between the "I" and the electrons.[20]

Charon's idea is novel. Electrons carry the soul or, as he puts it, the mind-spirit. But accounting for the vacuum adds new insight. If the material electrons are the soul-carriers, then those electrons that manifested, as it were, into positive mass-energy, would be concerned with communicating with each other about the physical massive world we all take to be real and "out there." The other electrons, those that exist under the Dirac Sea, the ones with negative energies, would consequently communicate different messages. They would speak to the manifested electrons about nonmaterial concerns and perhaps in not-so-distinct voices—from long off, from other worlds. These would be the voices of heaven and hell—the worlds of our imagination. The question is, how do these negative energy *virtual* electrons communicate with those that have made it to the other side—the positive *real* electrons?

The answer is through quantum physical correlation—the very "force" that causes matter to cohere as a result of the uncertainty principle and, I believe, provides us with consciousness. We will look at how our souls talk to us from the emptiness of space in the next chapter.

CHAPTER 7

Quantum Evidence:
The Self and the Soul

A Hundred torrents rise
From the surge my soul within;
The heavens in glad surprise
Stand still to behold me spin.[1]
—Jelaluddin Rumi, 13th Century Anatolia

Do you know Magic? Can you utter the name of your soul and
bring yourself back to light? Can you speak your destiny, create
life for yourself from yourself, as Temu created Ra? From the
light of your works do you know who you are?
—Normandi Ellis, *Awakening Osiris: A New Translation*
of The Egyptian Book of the Dead

At dawn, the moon appeared
and swooped down from the sky to take a look at me.
Like a falcon hunting up a meal,
the moon grabbed me and away we went!
I looked for my self, but my self was gone:
in the moon, by grace, my body had become like soul.
Luminous, I journeyed on as a soul, but all I saw was
that moon——
until the mystery of Self and self was completely clear.
Nine types of heaven, nine vibrations, all mingled in
that moon;
and the boundaries of my being disappeared in the sea.

Waves broke.
Awareness rose again
and sent out a voice.
It always happens like this.
Sea turns on itself and foams:
with every foaming bit
another body, another being takes form.
And when the sea sends word,
each foaming body
melts immediately back to ocean breath.

Thanks to my beloved friend Shams 'l Haqq, Sun of Truth!
Without his strength,
I couldn't see the moon,
I couldn't become the sea.[2]

—Jelaluddin Rumi, "The Moon and the Sea" from
The Diwan of Shams-i-Tabriz

AN OLD CHINESE CURSE STATES, "May you live in interesting times." We do. We are faced with unprecedented problems of population growth, high technology, medical setbacks and breakthroughs, abortion, and a growing dissolution of family and social values, to mention a few. How are we to deal with these problems? Here we are at the end of a century and a millennium, and, I wonder, do we really know any more than our ancient forebears? Perhaps. But are we really any wiser than they were? Perhaps not.

Looking at our planet in terms of how people live, with or without modern technology, in various states of wealth and deprivation, is like looking at a mosaic. There are clear areas of "haves" and "have-nots." We are technologically advanced here in the West, and I personally find this a great blessing. But there is still so much misery on this floating piece of sod that I often feel disheartened, especially as I look at how futile our attempts to right the world's wrongs are. I wonder just how science, at times the noblest attempt to deal with truth and humanity's problems, or spirituality is going to help. I believe science can help if we begin to ask the right questions.

Toward this goal, in this chapter I present a new vision of the soul and its relationship with the self. I sincerely believe this model provides a bridge, a new metaphor, connecting science and spirituality. It attempts to provide a new direction for inquiry, one that is not based on the misdirection resulting from following the ancient Greek path leading to reductionism, materialism, and passive epistemology as a basis for spirituality.

Up until now, science has been content with reducing all phenomena to the simplest processes. Anything complex had to be based on simpler material processes. Knowledge of the process remained passive. By learning about a physical system, no change ever took place in that system, nor was any change expected. With quantum physics all that changed. Extracting knowledge, learning about a system actually alters the physical system being examined in subtle ways. From quantum physics we learn that knowledge and matter are intertwined—changing one of the partners affects the other irreversibly. This fact provides a foundation for what I call *the new physics of the soul.* Armed with this knowledge, we will be equipped to move this world and its population to a new and hopeful destiny.

O, Keep Not My Soul in Cyberspace!

Physicist Tipler's model presented in chapter 5 is quite ingenious and a reminder to humanity that we need to take action in the universe at large. Assuredly, life on the earth must end in the quite finite period of seven billion years from now when the sun, going into its red giant phase, burns the earth to a crisp. I am not warmed by this vision.

I am also concerned with the computer-cold and soulless image proposed by the Dalai Lama. Is this all there is to being a human being? Namely, as put by His Holiness, *if all the external conditions and the karmic action were there, a stream of consciousness might actually enter into a computer.* Although the scientist in me is excited by the idea, my sense of sacredness and spirituality certainly is not. If life is somehow not that precious—and it seems not to be if I can find myself existing as a computer program in lifeless technocrystals of sand—then why all the bother?

A New Vision of the Soul?

It doesn't seem to help one bit to look at computer science or Buddhist thought in this regard. The Dalai Lama tells us that our souls may be able to inhabit artificially-intelligent bits of silicon dioxide, and that seems strange, indeed (even more perplexing as we shall see in part 3). In Buddhism we face the face we have before we are born! The paradoxes of Buddhist logic lead to the conclusion that none of us even has a soul—the soul is ultimately an illusion, a wishful thought. I paradoxically conclude from this that while I presently enjoy the blissful notion of my soul, the *I* that enjoys it is an illusion and does not exist! *I* am *not-I*, even if I think I am!

Something seems to be missing from all of this technologically created *nonsouldom*. For nowhere do I find *myself*, and although this self of mine may not be precious to you, it certainly is to me. Even scientifically speaking, somehow it seems to be a great leap of faith to think that a computer program can do anything like create a "me" with complex rapid simulation. Suppose a computer program *can* generate a soul with no more difficulty than numerous, ingenious, and complex operations performed in semiconducting sand: What can we gain from this new vision?

I doubt that a computer program can generate a soul. But it turns out that a grasp of computer automata—devices that are capable of "observing" and storing memory—leads to an essential insight into how the soul communicates with the self. With this insight we can see an essential difference between the self and the soul, or, as put by the poet Rumi, the *Self* and the *self*. The little *self* disappears when the moon grabs him and reveals his big *Self* to himself. Rumi, in poetic metaphor, tells us there is a difference: *Self*, a reality, appears to contain *self*, an illusion. That difference depends on information: not only what information consists of, but who or what has it. To see how this difference arises, we begin with the role that consciousness plays with matter.

Are Human Beings Conscious?

Materialistic scientists—those who believe consciousness and soul are no more than epiphenomena arising from complex interactions in matter—despite the obvious subjective fact of their own awareness, often ask the above question in attempting to discern what is required for consciousness to be present. Recently, the validity of this question was put to the test. In April of 1994, I was invited to speak to the first-ever conference "On the Scientific Basis for the Study of Consciousness" held at the Medical School of the University of Arizona, in Tucson. A number of well-known scientists and brain researchers had gathered to decide once and for all: Is it scientifically provable that we humans are conscious entities?

You might laugh at this seemingly stupid question. Of course humans are conscious: just look around. You can see them walking about, talking to each other, and apparently making sense of the world, thank you. What could you mean by such a question? What would need to be proved? That's just it. We scientists don't know how to prove the existence of consciousness in a demonstrably scientific manner. To be honest, we aren't really sure if consciousness can be measured or if it even exists in the world of space and time.

The big problem at the conference was put concisely: Scientists simply don't know what consciousness *is* and they don't quite know where to look for a seat of consciousness. Where is the soul, or, is there one to look for!? Is the soul in the body? Well, we know a lot about the brain, the nervous system, and how the body functions when the brain is awake, asleep, or under anesthesia—all topics of the conference. We know when a person is unconscious, and we know what the brain is doing when a person is thinking about an apple or lusting after someone. But when it all boils down to proving something, to demonstrating the existence of consciousness residing somewhere, we simply can't find any scientific evidence that a place of consciousness exists!

How strange! Here we are approaching the end of a millennium, embarking on a science of consciousness, and we can't even come to any scientific agreement as to just what substance consciousness consists of. Is it something physical? If it isn't physical, then how does science deal with it?

Leaving that aside for the moment, we certainly are no better off attempting to define the soul as a seat of consciousness or as an entity. There may even be some argument about whether or not the soul, if it exists, is even conscious. You may wonder about that, but if the soul is not physical, and consciousness requires a physical substrate in order to exist, then how could a nonphysical entity be conscious? Well, suppose first of all that the soul indeed has substance, i.e., carries weight. Then what?

The Soul Weighs In

Dr. David E. H. Jones, a professor at the University of Newcastle-upon-Tyne better known to British readers by his pen name, Daedalus, made a seemingly outrageous proposal regarding the physical existence of the human soul.[3] Over several years, Jones, as his alter ego Daedalus, has proposed perfectly sound ideas from science blended with outrageous humor. His writings grace the weekly pages of the prestigious British scientific journal *Nature*.

What Jones proposes has been considered before. Namely, if the soul is a physical system within the body, shouldn't it have physical substance? In other words, at the time of death, does the substance of the soul leave the physical body as a physical ejection? If so, could a delicately balanced physical measuring apparatus measure the departure of the soul? For example, if the soul has weight, would a scale balance measure the change in weight at the instant of death (assuming that death had an instant)?

Remember that Aristotle (chapter 2) based his vision of the soul on it being substantial, but of a very refined essence, something feathery, like the finest gossamer, in constitution. Also for Aristotle, the soul was, and I stress this point, immovable substance. Now the question is, could such substance have any weight?

The weighing of souls idea is rooted in ancient history. As we saw in chapter 4, the early Egyptian underworld gods, Anubis and Thoth, balanced the ba or heart-soul of the expired scribe, Ani, against Maat's feather of truth. If his heart was too heavy, it would be devoured by a waiting monster. Jones proposes to use sensitive piezoelectric crystals as transducers, possibly connected to inertial navigational acceleration-detectors. Presumably when the soul departs the crystals would register a vibration.

Daedalus suggests that resolution of soul-weight could end the debate about abortion. By applying the soul detector to pregnant women, investigators could check theologians who argue that the soul enters the embryo a week or so after conception. "It is clearly worthwhile," he writes, "to establish this moment accurately. If the soul turns out to enter the fetus quite late in pregnancy, the religious arguments against contraception and early abortion will be neatly disproved."

Taking Your Soul for a Spin

Even though a scientist, more than likely jokingly, says he can measure our spiritual selves through our loss of body weight at death, there is little evidence that the soul possesses physical density. Although I seriously doubt that the soul has weight—in fact, I see no reason for it to have any mass at all—another physical possibility, also raised by Daedalus, may have scientific merit.

If the soul corresponds to a system of spinning particles (all fundamental particles posses an attribute called spin, a quantum property that determines the particle's interactions with magnetic fields and with other particles that also have spin), then a sensitive magnetic field measuring device might detect the spin of the soul when it left the body. For example, by placing the body in a magnetic resonance imaging device, a large piece of equipment commonly found in modern hospitals and used in diagnosing tumors, it may indeed be possible to detect the exact moment that the soul passed from the body. But now we have another question: Why would the soul have a spin?

"Traditional theology is silent on the spin of the soul," Daedalus wryly comments, "though it may predict that the soul of a sinner would depart downward, and might weigh less than that of a righteous believer."

Amusing though it be, the idea of a spinning soul may explain how the soul interacts and communicates with the particles making up the body. In fact, if the soul is composed of spinning fundamental particles, that fact alone helps us to resolve two ancient philosophical mysteries:

1. The mysterious heartfelt relationship of one soul with another, particularly in dealing with the question of soulmates;

2. The mysterious relationship of the soul with the self, including its identification with the body.

The solution to mystery 1 comes through understanding how each person loses awareness of space and time. When two beings have a heartfelt relationship with each other, they each lose a sense of location. To some extent, each one feels the other's presence as if each was the other as well the one. The solution to mystery 2 is similar, only in this case while the one is in space and time, the other is not. The relationship therefore takes on a different quality: body and soul.

A Soul-Spinning Tale of Heaven and Earth

We shall explore the two mysteries mentioned above by looking metaphorically and physically at the possibility of a soul—made up of spinning "virtual" negative energy electrons embedded in the vacuum, interacting with a body—made up of real positive energy electrons moving around in every living cell. At the end of chapter 6, we examined how negative potential energy electrons, those immersed in the vacuum of space known as the Dirac Sea, could communicate with real electrons within our bodies and came to a conclusion similar to physicists Jean Charon and Shiuji Inomata that the Dirac Sea alters our thoughts and feelings as much as it buffets matter about.

Here, using an imaginative quantum-physical model employing spins of particles, we examine how soul-to-soul and soul-to-body communication could take place. I want to see how far I can take the spinning-particle-as-pieces-of-the-soul metaphor. Assume, at the risk of sounding reductionistic, that the soul is composed of billions upon billions of small spinning negative energy particles. Consider that each of these "virtual" particles is embedded in the empty vacuum of space as I discussed in the previous chapter.[4] That's a big assumption, but please bear with me. Imagine that each virtual particle interacts with one other real particle, an exact duplicate—another spinning particle within the body.[5]

This virtual "heavenly" particle has qualities (spin, negative energy, and mass) which mean it is *potentially* able to be a real particle, but it isn't as long as it remains a virtual particle. Remember, only when such a virtual particle "pops" out of the vacuum does it become a real particle. Then it leaves behind a hole, an antimatter particle with opposite characteristics.[6] Here we shall leave it in the Dirac Sea.

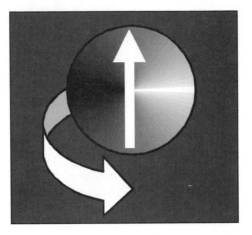

Figure 7.1. The Spinner: A Particle That
Makes Up the Soul's Mind and the Body's
Matter. The curved arrow shows the axis of
the spin.

Nevertheless, even though one of these spinners is in heaven (the sea) and the other on earth (the body), they have continued contact—a long and intimate encounter. What does quantum physics have to say about this virtual particle/real particle encounter?

Some Basic Pointers About Spinning Particle Physics

Before I tell you about the virtual particle/real particle encounter, I need to provide a few basic facts about the physics of spinning particles. Particles are grouped in families according to what are called quantum numbers. These numbers refer to specific observable features of the particles, called observables by physicists. Thus to every observable there corresponds a specific quantum number. Spin is an observable and, as we have been able to observe, all fundamental particles have it. Spinning particles can be pictured as tiny little tops with their axes of spin pointing in specific directions of space. So they have what are called *spin quantum numbers*, and these numbers can take on increasing integer or half-integer values. The lowest spin value of a particle is ½.

A *spin-½* particle can only be observed pointing up or down with respect to an apparatus reference direction.[7]

When two particles with spin-½ interact with each other with sufficiently low energy,[8] their composite spin, taken to be the total spin quantum number when both spins are added together, must be zero ($[+½] + [-½] = 0$). This

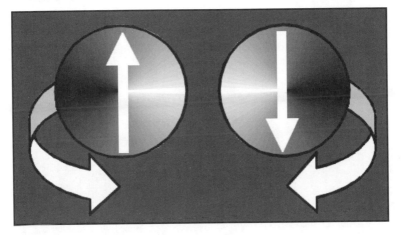

Figure 7.2. Two Spinners. After interacting, if one is pointing up, the other must point down.

would seem to mean that one of the particles has its spin axis pointing up and the other down. But this isn't the end of the story. According to quantum physics, although their composite spin is always zero, the direction of each particle's spin is not determined. According to the parallel *mind-possibilities* model of quantum physics,[9] each particle, after the interaction, simultaneously spins in two opposite directions![10] This remains true so long as no one observes either particle. Each spin-direction represents a mind-possibility.

Back to a Yarn-Spinning Encounter

Now let us look at a body/soul real particle (labeled *A*)/ virtual particle (labeled *B*) encounter to see how this addition to the story plays a major role. Accounting for their spins pointing in opposite directions to reach a total of spin-0, and the two different mind-possibilities, we have particle *A* with *spin-up* and particle *B* with *spin-down* or in the other mind-possibility, where the spins are reversed, *A* with *spin-down* and *B* with *spin-up*.

Let us for the moment concentrate on one of these particles (say *A*) and ask about its status in each of these mind-possibilities. In mind-possibility 1, *A* points up and the virtual particle, *B*, points down. In mind-possibility 2, the reverse is true.

So far I am well within quantum physics principles. Now I want to go outside of those principles by giving each particle, virtual or real, the ability to hold a memory—bits of data consisting of spin quantum numbers. What would *A* have in its memory if it could observe its memory? Surprisingly, it would "see"

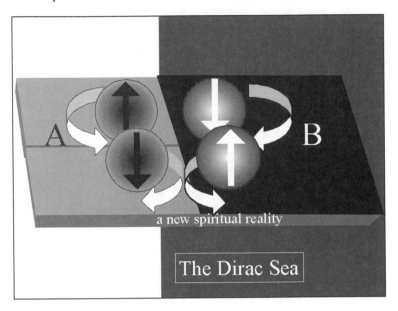

Figure 7.3. Heaven and Earth Parallel Minds Model. A spinning virtual particle (B) interacts with a duplicate but real particle (A).

itself as the overlapping whole system of *A* and *B*, and not just as *A*.[11] By *remembering*[12] itself as a part of the overlapping spin-0 system consisting of itself and the virtual particle, it no longer "sees" its individual state as simply particle *A*. This remains true no matter how far apart the two are in each mind-possibility.[13] The individual particle *A* no longer "thinks" of itself as *A* alone, but only as the *A-B* combination. *A*, in essence, spreads over space in each mind-possibility as far as *A* and *B* are separated. *A* "thinks" of itself as being that size as it "views" itself in this strange overlapping mind-possibilities situation.[14]

Now let's see how this overlapping of possibilities solves our two soul mysteries.

The Mysterious Heartfelt Relationship of One Soul to Another, Particularly in Dealing with the Question of Soulmates

If the virtual particle B were a real particle, we would call this the particle-mate model. Then we could think of the pair as the joining of earthly particles in a particle-mate situation. Even after the particle interacts with the particle-mate, the particle continues to "see" itself as both, particularly since *A* and *B* are *identical*.[15] This is what is meant by gaining one's soul and losing

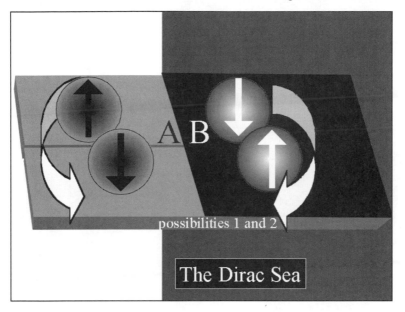

possibilities 1 and 2

The Dirac Sea

Figure 7.4. An Overlapping of Possibilities. The two are pointing both up and down at the same time, giving spin-0.

the world. Specifically what is lost is the sense of place or location. When this loss occurs, the two are as one; there is no apparent separation, and yet they are physically separated. This explains mystery 1.

The Mysterious Relationship of the Soul with the Self, Including Identification with the Body

We could also imagine the situation shown in figure 7.5 to be representative of the heavenly virtual particle and the earthly body particle in interaction. As we shall see, this will lead to explaining mystery 2. However, there is a subtle difference here between mystery 1 and mystery 2, namely, this notion of location in space and apparent reality of one of the particles. In 1 both particles are earthly. In 2 only one of them is. But, don't ask which one. Since A and B are identical, even though one of them is earthly, so to speak, there is no way to know which is which or which is where. Such questions belong to the realm of the mind.

When it comes to identical particles that are interacting or have had a past interaction, there is no way to determine these seemingly innocuous appropriate labels. Such is the way of quantum things. Such is the rule that

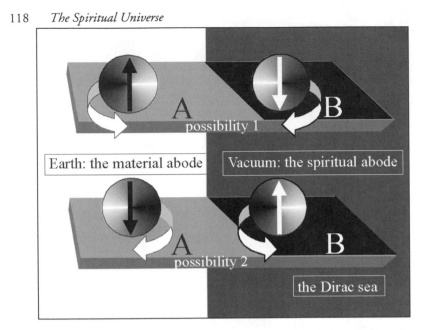

Figure 7.5. A New Spiritual Reality: The Virtual Particle and the Real Particle in Interaction. After interacting, the real and virtual particles form a bond. Don't ask which is which. Together they compose a single spiritual action.

quantum things must obey. While this loss of spatial identity was also true in mystery 1 as well, both particles were real. This led to a dizziness, a loss of spatial sense, a psychic awareness, a sense of connection with another physical thing. In other words, falling in love. But in 2, *A* has no one else to refer to. Its dizziness, its falling in love, its connection cannot be with another physical thing like itself. There is no one there, no one "out there." Its connection is felt as ghostly, mystical, spiritual, love of God, or the like.

There is something else mysterious in this heaven-earth match.

Really One in Two Places at the Same Time

The otherworldly aspect of this heaven-earth, parallel mind-possibilities interpretation of quantum physics makes things strange indeed. The two spinners are separated in virtual/real (V/R) space in each separate mind-possibility, and yet they are connected in V/R-space by their being in two mind-possibilities simultaneously. If either of the mind-possibilities were to vanish suddenly, the two would no longer "feel" any connection in the V/R-space in the remaining mind-possibility.[16]

a new loving reality

Figure 7.6. A New Loving Reality. Joining the real particle with its real partner-particle, we have a match.

We might adopt Plato's idealism for the moment and picture the two mind-possibilities accordingly. For mind-possibility 1, mind-possibility 2 is ideal and for mind-possibility 2, mind-possibility 1 is ideal. So long as no attempt is made to "know" which mind-possibility either particle is "really" in (whatever that means), the pair only know of themselves as the one and only one particle that is. This one entity is not even knowledgeable about its existence in two mind-possibilities at all. It just knows spin-0. It just knows it *is* and this knowledge exists as a unified state of awareness with no knowledge of separation into two, *A* and *B*.

Forgetting the Soul

But now something else can happen. Instead of remembering itself as the larger spin-0 system, as the "over-soul," so to speak, it can remember itself separately and, in so doing, lose its soul-identity. This action completely separates it from its virtual-reality partner.[17] We could look at this as mind entering matter. Most important, this action of separation from the whole is the *primary action of higher or soul-consciousness*. When this observational action occurs, the "over-particle" separates into a virtual particle and a real particle. This self-remembering action separates the real from the imagined and introduces two notions of time: imaginary time and real time. One might see this as V/R-spacetime turning into separate virtual spacetime and real spacetime. This separation is the first step of spirit entering into the process of *embodiment*, it constitutes the fundamental basis for awareness of the body as a feeling of loss of soul.

When this separation happens to previously embodied real particles, a feeling of body awareness arises as loss of another.

Remembering the Soul

All is not lost. Redemption is at hand. The act of remembering can be achieved (1) *without*, as discussed above, or (2) *with* knowledge of self in separate mind-possibilities. Accordingly, this knowledge can be gained in two different modes: The real particle either . . .

1. Sees itself only as the part while forgetting itself as the whole; or
2. Sees itself as the part while remembering itself as the whole.

In 1 the real particle knows itself, but it has interacted with other matter and hence *told someone what it knows*; it no longer remembers that it is part of a larger whole. I call this *forgetting the soul*. We can think of this as the fall of the soul into matter, or soul-loss. The "telling" separation is quite important. Here it loses one of the mind-possibilities, gains self-knowledge, and knowledge of only one mind-possibility.

In 2 the real particle knows itself and knows it is part of a larger whole simultaneously but, paradoxically, only so long as it doesn't tell anyone else, even itself. (I'll explain this in greater detail in the remainder of the chapter.) I call this *remembering the soul*. We can think of this also as the material part of reality rising to reach the spiritual part of virtual reality. Here the real particle realizes itself in two mind-possibilities and senses its connection with its heavenly host or, if with another physical entity, an earthly partner it has previously interacted with. This is the *secondary action of higher or soul-consciousness*.

Each action arises by how a question is put, i.e., how the act of remembering is carried out and what is done with the information obtained.

The New Physics of the Soul

From the simple model presented above, we can see how questions and answers put to an intelligent system, a sentient being (a being with a memory), alter the physical processes occurring within that being. Now we shall see why following the Greek form of reasoning based on either idealistic or realistic models of the soul misled us. The attempt to model the soul-body interaction in terms of knowledge defines what I call the new physics of the soul and demarks the beginning of the difference between a self and a soul.

Let me briefly summarize. By dealing with virtual particles in the vacuum and real particles in a sentient body, I imagine the soul as composed of spinning virtual, negative potential energy, electrons that I called virtual particles immersed in the Dirac Sea-like vacuum of space. Each virtual particle has a memory and interacts with real spinning particles in the body. This model solves two mysteries dealing with how the soul communicates with the body. In solving mystery 1, the heartfelt relationship of one soul with another, a simple quantum physics model shows how beings communicate at the soul-level and sense a loss of self. In solving mystery 2, the relationship of the soul with the body-oriented self including identification with the body, the model shows how the heavenly virtual particle and the earthly body particle interact. Here the self is part of a *sacred* or spiritual whole. In both cases the communication involves a new quantum physics relationship of knowledge with matter.

Before I go into this in greater detail, let me explain something else about the nature of *knowledge* and *being* when quantum physics enters the metaphysics.

Knowledge and Being

We saw how the soul, losing its *heavenly* awareness, can believe it is located in the body through an action called *self-knowledge*. Next I want to show you how the very act of learning something objectively actually *extracts* some form of embodied or *secret* knowledge/essence from matter. Nothing could be farther from Cartesian dualism.

Cogito ergo sum, Descartes announced and in so doing fused together two contradictory thoughts. I think *(cogito)*, Descartes takes as absolute fact. Therefore, I am *(ergo sum)* follows as a deduction. This means existence of a body results from thinking. Thus, a body cannot exist if it isn't thought about. Or, mind is primal and the body is a secondary thing arising from thought. This reasoning contradicts what Descartes says next: "I am a *thing* which thinks." He is implying here that there are things that think and things that don't think. From this, Cartesian dualism, as implied by Plato's idealism, follows: mind and body or soul and body are separate things.

Our present thinking is dualistic, implying mind and matter are separate and cannot influence each other.[18] Such thinking occurred before Descartes's famous cogito dictum. In ancient Greek metaphysics, in fact up to the beginning of the quantum age, being and knowing were separate things, like apples and oranges. You should not confuse one for the other, as if they were inter-changeable. In the popular vernacular, don't confuse the map with the territory. The map (knowledge) and the territory (being/existence) are very different.

But with quantum physics, the *principle of uncertainty* muddies the wa-ters and blurs the map. What we know and how we come to know it alters the thing we seek knowledge about. The act of knowing is always accompanied by an action in the physical world. So gaining knowledge about a thing always means, in some manner, physically altering the thing. Knowledge can be imagined as being pulled out of the thing. Somehow the very act of learning something extracts some form of knowledge/essence from the thing.

Knowledge, however, is not material. At a minimum, it is not matter in the same sense that inert and dead stuff is matter. It may be a fine form of matter as Aristotle thought the soul was, a subtle form of activity. Even though knowledge is not matter, it, nevertheless, affects matter in much the same way that a field-of-force affects matter. Just as a field-of-force (for exam-ple, an electromagnetic field) surrounds every bit of electrically charged mat-ter, a field-of-knowledge can be pictured as surrounding matter, too. We then come to see knowledge as something that exists in space and is capable of per-sisting over time. When knowledge changes, things also change.

The Morphosis of Knowledge

Next we see that the possession or encapsulation of some information, hold-ing it in secret, violates the *uncertainty principle* and provides soul-awareness, while sharing it creates soul-forgetting and the familiar sense of self within the body. Thus the embodied self arises when it communicates soul-knowledge with the world-at-large. When soul-knowledge is communicated, the soul falls into the body—the connection with the vacuum sea is temporarily bro-ken. Something new is born. A child emerges. A sense of self manifests. One awakens. Matter becomes conscious and perhaps, paradoxically, the connec-tion to the soul is lost.

When soul-knowledge is held secret so the bond between the heavenly partner (the vacuum virtual particle) and the earthly entity (the real particle) remains intact, the soul knows no separation. Each particle knows no subject-object distinction. Each one knows no *thing*. For Descartes, this would be im-possible. Each particle has the experience of knowing without an object being truly present. This state or characteristic of knowledge and consciousness is difficult to explain. In this state life and death are one. Being and knowledge are one. The unborn and the born and the never-born are one. The question is: Is it possible for a thing to have this knowledge, secret knowledge as I mentioned earlier, that remains with the thing and is therefore not extracted from it, and, simultaneously, be in the world?

The possession of such secret information, it turns out, violates the uncertainty principle, so one would gather from this that such secret knowledge cannot be held. However, there is a subtle but important distinction between a thing holding information about itself and holding information about the outside world. Knowledge of things outside of the self must be contained in memory. Such knowledge necessarily must be fuzzy because it consists of information gained by the self interacting with the outside world. As such, this information cannot violate the uncertainty principle. Indeed it was through such interactions that physicists first came to realize that there even was an uncertainty principle. But because it remains subjective,[19] secret knowledge or soul-knowledge and its achievement called soul-awareness can occur with violation of the principle of uncertainty. Only when this knowledge is extracted[20] does the force of the uncertainty principle come into effect, blurring the edges of objective definition while, nevertheless, providing objectivity. This result, a fairly new discovery in quantum physics,[21] has repercussions physicists are still looking at. The physics of the soul can be stated thus:

> Soul-knowledge held maintains the soul-to-self connection. Soul-knowledge extracted creates identity of entity within space and time. In other words, when the soul falls into the world and takes form, it learns, but it gets lost.

Figure 7.7. The Fall of the Soul. When the soul falls into matter, it gains a self.

Self Identification:
The Fall of the Soul into Space-Time

Now we explore the processes metaphorically referred to as the fall of the soul from heaven and the rise of the self from earth to realize the heavenly soul. Let's see how all this happened by returning to our spinning-particle model. Only this time, I'll use faces instead of spinning spheres to represent the self and the soul. First we will look at the fall. It takes place in three stages:

1. When the particle contains only knowledge of total spin-0, it has only soul-awareness. It does not know where it is, what it is, or when it is. It is not in heaven, hell, or on the earthly plane. It, simply, is aware. This is shown in figure 7.8.

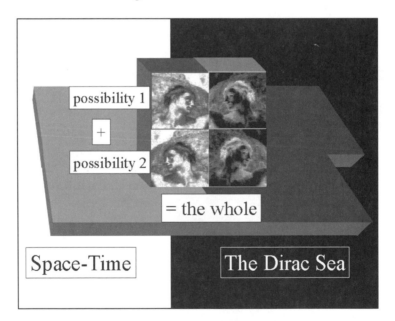

Figure 7.8. Stage 1: The Self and the Soul as an Unbroken Whole. As in the two-particle spin model the self and the soul are taken to be two particles, each capable of being in two states. Instead of *spin-up* and *spin-down*, their faces are either toward or away from each other in each possibility. Here they form an unbroken whole, an overlapping of possibilities. When they are in this dialogue each knows not space-time nor vacuum (Dirac Sea) order. Here both possibilities are present, but neither the soul nor the self has completely identified itself to itself.

2. If it now asks about where it is and in which direction its spin-axis
   points, but does not attempt to extract that knowledge from itself
   and give it to the outside world,[22] then it, paradoxically, learns its total spin—its soul-to-body connection—and it becomes aware of its
   own spin direction, that is, its material body. Here it contains subjective knowledge (it knows its own spin and that it is part of a two
   particle spin system) which violates the uncertainty principle. Consequently, it is partially in space and time and partially out of it. This
   is shown in figure 7.9.

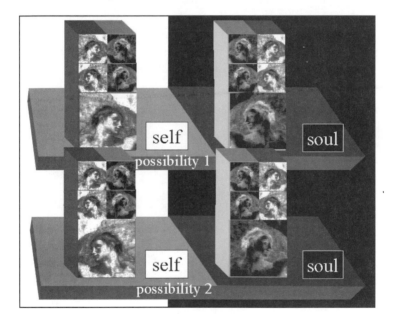

Figure 7.9. Stage 2: The Self and the Soul as a Nearly Broken
Whole. The self and the soul in each mind-possibility. Each knows its
own face, in which direction it is looking, and each one knows or remembers the whole from which it now feels some separation. The self
has taken a stance and partially entered into space-time. In so doing,
the self has lost knowledge of the other possibility except through the
memory of its connection to the soul. This is a critical stage, for the
self can further disconnect from the soul or it can return to stage 1.

3. If it now attempts to fully orientate itself in space and time by sharing its, now, recently acquired knowledge of its spin direction with
   the outside world, it loses its connection with its heavenly partner
   altogether. Now it knows itself as reflected by the outside world, and

consequently its knowledge is utterly subject to the uncertainty principle. It actualizes the fall and its knowledge is completely contained in matter. This is called self-identification and is shown in figure 7.10.

To do 2, it must know with certainty that it has spin of its own and it also must know, secretly, that it exists as a spin-0 system. We might call this the double-awareness of body and soul or the connection of the virtual particle with space-time-matter. As long as the spin-0 knowledge is kept secret, even though the virtual particle has made contact with the real particle, the soul has not gotten lost. However, the soul can get lost. It can fall into space and time.

To do 3, it must know with certainty that it has spin of its own and forget that it ever existed as a spin-0 system. It loses the double-awareness of body and soul or the connection of the virtual particle with space-time-matter. It has fallen into space and time.

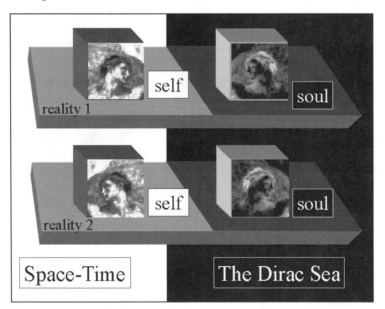

Figure 7.10. Stage 3: The Self and the Soul Are Separated. The self and the soul are now completely apart in each mind-possibility. Each only knows its own face, in which direction it is looking, and now feels absolute separation. By not remembering the other possibility, the self knows its face and has entered into space-time. In so doing, the self has totally lost knowledge of the other. This is the final stage, for the self is disconnected from the soul.

How You Can Lose Your Soul

From the above three-stage model we grasp what is meant by soul-loss, a common event in the shamanic world. Soul-loss occurs when the entity no longer holds its soul-knowledge in secret, which in human terms results in the loss of sacred knowledge and the fall of the soul into spacetime: "loose lips sink ships," as the old World War II adage goes.

Of course, the entity gains something in the loss. It learns that it is a self. It identifies itself as real particle *A* and not virtual particle *B*, but it loses contact with *B* because that connection involved each particle holding in memory two mind-possibilities. The particle knows that its spin is pointing upward, for example. It has shared this knowledge openly with the world-at-large. It has fallen into a subtle trap called *ego*—the sense of embodied self. It learns that it exists as an entity with a separate *id* in a single world. In other words, it becomes an *id-entity* (an identity) when it loses its connection to its virtual partner. In fact, once this happens, it has no memory of such a connection at all, or at best it is uncertain whether it ever had such a connection. This identification with ego, the forming of id-entity arising from the sharing of its individual *A* state with the world-at-large, is called soul-loss.

Soul-loss occurs when the entity *A* no longer holds its soul-knowledge as secret, which in human terms results in the loss of the only sacred knowledge it knows. "Secret" not only sounds like "sacred" but has a similar meaning in its definition as "beyond ordinary understanding, mysterious."

With the arrival at stage 3 and the fall into matter, this soul-loss action of identifying projects knowledge from the spin-0 pair into the world-at-large that originally encompassed the two mind-possibilities. In forming an id-entity, in extracting knowledge, the world-at-large demands its pound of flesh. It creates uncertainty; even stronger, it demands its creation. No longer can the two, *A* and *B*, be known as a spin-0 pair. No longer is there a bond of essence, involving the overlap of two mind-possibilities, between the two. The pair knows separation for the first time. And each is alone: the virtual particle in heaven and the particle on earth.

Can this seemingly hopeless situation be remedied? Can the virtual particle's presence be remembered by *A*, knowing that it is part of a bigger picture, the spin-0 system while simultaneously knowing that it is a body? Can it be an entity and not lose its soul? Can it have simultaneous knowledge of heaven (vacuum) and earth?[23] In other words, how does it return to stage 2?

The Return to Secret Knowledge: Remembering the Sacred

While sharing knowledge leads to the separation of soul and self, here we shall see that an entity can be in the world and not lose its soul. It does this by remembering, literally putting its members back into contact. It can, provided it doesn't tell anyone about it when it does it. It must keep this as secret knowledge, possibly even from itself! By remembering itself in each mind-possibility, it can reconnect the simultaneous knowledge of its own spin state (*up* in mind-possibility 1 and *down* in mind-possibility 2) and also the virtual particle's spin direction, with the spin-0 system. However, it can't communicate that knowledge to the outside world-at-large without disrupting its own secret/sacred-knowing state. It may not even tell itself what it knows. In practice, this action of remembering becomes an *act of faith*. Thus arises, if you will, a *physics of faith*. Faith implies knowing without proof.

So it is this action of *soul-remembrance* that defines self and soul conjunction. In this case the remembering entity possessing *A-B* knowledge sacredly and secretly id-entifies with both the larger over-particle *A-B*, and the smaller body *A*. If that knowledge were extracted through interaction with another system in space-time by proclaiming its id-entity as *A*, the system *A-B* would fall into knowledge of *A* alone, causing loss of soul.[24]

The Nature of Prying

If we were to include an interaction with another particle *C* so that *A-B-C* comes into existence, the over-particle would grow even larger,[25] resulting in a sacred bond between the three entities. But when the triplet inquires into its own nature, asks about its parts, so to speak, and each identifies with either *A*, *B*, or *C*, it must do so carefully or else it can lose its over-connection to the others. For an entity to do this, to hold itself in stage 2, it must be and remain conscious. What the individual particles know secretly separates them from all other particles with whom they do not share such knowledge. But if the *A-B-C* system communicates with the world-at-large, if the system enters stage 3, it loses its connection to its various parts and each part falls into identity and loses consciousness. Here we see how knowledge can actually cause loss of awareness.

The Communion of Spinners

Next, we will see how a third observer, *C*, can recognize either the connection of the first entity, *A*, with its soul, *B*, but not its specific identity, *A*, or it can recognize *A* alone and in doing so causes *A* to lose its "soul-connection" with

B. Now what do I mean by soul-connection? A soul-connection exists when two particles are connected by the existence of overlapping possibilities as shown, for example, in figure 7.4.

Returning to our two partner *A-B* system for the moment, suppose this other system *C* gets nosy while our original system, *A-B*, is meditating on the state of the universe and such. If *C* attempts to extract knowledge from this system, two possibilities can occur. If the inquiry attempts to extract the knowledge state of the total system of *A-B*, i.e., the knowledge that the total is spin-0, no problem![26] In fact, in so doing, *C* joins with the original system by acknowledging the secret knowledge within the original system. You could call this *communion*.[27] However, *C* doesn't know *A*'s state. But, *C* could find out. Now, the mere asking of the wrong question, a question about *A*'s spin direction, will destroy the spin-0 state. If *A* attempts to answer *C* and thereby share its sacred, secret-knowledge with *C*, *A* will no longer remember itself as part of the larger system.[28]

This consideration of changing the state of a system by the mere response to a question bears on sacred mythology as well as metaphysics. In fact, we may be looking at the quantum physical equivalent of Adam and Eve's notorious exodus from the Garden of Eden when they ate from the tree of knowledge.

Adam and Eve: The Quantum Physics of Knowledge and Identity

The sharing of the three-way knowledge between God and Adam and Eve leads to some interesting consequences if one takes a God's-eye view regarding the sexuality of the soul and the masculine and feminine roles that God plays.

According to Judaism, Adam and Eve had it made, with plenty to eat, a life full of goodness and the absence of evil. Or did they have it so good? For if one looks carefully at the old tree in the garden, it bore the fruit of the knowledge of good and evil. Thus, before Adam and Eve ate that fruit, they had no knowledge of such things. Things, events, observations, everything they saw and did carried no taint of evil, glistening of good, or *distinction from God*. To couch it in the terms of quantum physics, at first Adam and God were connected, as in the spin-0 state. Then God tells Adam who he is. This corresponds to the secret soul-knowledge that Adam knows who he is and knows simultaneously that he is part of the God-Adam connection.[29]

Then, Eve then enters and forms a sacred bond with the other two. She knows secretly that she is Eve and that she is connected to Adam and God,

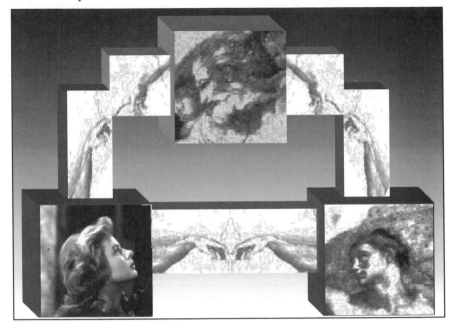

Figure 7.11. A God's-Eye View of the Whole. Here God, Adam, and Eve are seen as a connected whole.

but she doesn't know who is who. She cannot distinguish God from Adam, for to do so she would break the bond between all of them. Once Eve is present, the same is true for Adam; he can only know Eve and God as the over-soul and is unable to distinguish between them. In fact, it would not occur to Adam even to make the distinction. Nor would it occur to Eve. If I may be so bold, the same is true for God, who lives in virtual reality in the vacuum of space. Their state is simultaneously a trinity and an identity. Each secretly knows separately an id-entity and collectively a soul.[30]

This sharing of the trinity knowledge between God, Adam, and Eve leads to some interesting consequences if I may take a God's-eye view. For example, when God recognizes Himself, he sees Adam. When God recognizes Herself, She sees Eve. When God recognizes humanity, They[31] see Adam and Eve. Adam in seeing Eve sacredly, sees God and Eve. Eve, in seeing Adam sacredly, sees God and Adam. Herein lies a clue to the sexuality ascribed to God: each sex sees God as the other. For a man, God is symbolized by a woman and vice versa for a woman. This may also be the quantum physics behind Jung's model of *anima* and *animus*.

When Adam and Eve ate the fruit, the sacred/secret knowledge entered into consciousness, and they knew separation from God for the first time. All three separated. All entered into stage 3. God, or the oversoul of Adam and Eve, separated from them. They separated from each other, and Eve knew herself as naked in the garden. And, I would add, they became self-conscious for the first time as they also gained id-entity of themselves as separate from their virtual selves in heaven or from God.[32]

Secret Knowledge and Soul Connection

Let me reiterate: secret knowledge identifies you and separates you from all others, but not from your soul. Sharing that knowledge, which in the ancient texts would be called profaning that knowledge, causes you to lose your soul-connection and helps others along the way to losing theirs. Thus, we see that quantum physics uniquely offers a new model for self, soul, and oversoul. As far as I know, no philosophers have thought of this before. With the one exception of physicist-philosopher David Z. Albert, no others have dealt with the peculiar physical properties that arise in quantum mechanics from the measurements associated with knowledge and identity.[33]

Allow me to assume, as I am sure you have surmised by now, that what I described about the particle/virtual particle applies at the level of the human soul—the *sentience of spirit*. What enables you to identify with your body and keeps you apart from the outside world is not physical space or time, but knowledge. As long as you secretly know something about yourself that the outside world at large does not know, namely your connection to the oversoul, you are uniquely connected with your sacred spirit. This secret knowledge identifies you and separates you from all others, but not from your soul. Your existence as a separate being with individual characteristics arises with or without this knowledge. With the secret knowledge you are aware of your soul. Without this secret knowledge, you are not.

At the level of particles you and everyone else are completely the same. So in some sense your separation in mind or consciousness from anyone else is an illusion. (In chapter 12 we'll see how, at the level of consciousness, no distinction can even arise between you and anyone else.) Indeed, the separation you feel from everyone else is a mind or knowledge-based thing. We point to ourselves as singular, with unique fingerprints or noses or eye color, weight, height, sex, race, etc. But all of these attributes are somewhat artificial and come from the knowledge-base we have accepted as truth, and not from anything really fundamental.

The Western civilization problem of soul-loss, at the level of the spinning particle, arises in the form of asking the wrong question—the question of *what* separates us—which in turn *actually* separates us! The wrong self-knowledge acts in the world by creating divisions, space and time walls. I ask you to tell me your intimate secrets and I change you forever when you tell me. Tell something about myself to the world and I grow more separate from it. I hold the secret knowledge within myself and I merge closer to the universe, to the oversoul! This merger/separation is redefined every time I consider who I am and every time I accept answers from the outside world that identity me as some image and disrupt the internal and eternal harmony of my *soul and self* dialogue.

In this simple model several spiritual truths come to light. The old adage that sacred experience cannot be explained leads the way. Prayers are best done in silence. The spoken *Tao* is not the *Tao*. The concept of never saying *I* or giving oneself too much praise or credit for tasks accomplished also comes to the fore. How I share myself with the world creates my world. What we say to each other, and how we say it, is more important than we could ever have believed from a Newtonian, mechanical, or classical physics point of view. (I'll have more to say about these processes of knowledge-sharing and becoming in chapters 12 and 13.)

PART 3

THE RISING SOUL: INTO NOTHING WE GO

THE VACUUM IS THE strangest place we have ever been, if we can ever remember it. It's not a place really, and it's not a not-place either. It's real and it's not real simultaneously. It's everything, and it's nothing. So let's suppose the worst: we are nothing. We come from nothing. We return to nothing. So what do we do? We continue our quest to rebuild the soul, so to speak. We have seen that to know the soul means looking at a very ancient picture of it and juxtaposing that with the most modern view we can muster. In part 3, we continue this plan. Here we begin to reconstruct the soul and to find the right questions. You might say that part 3 has Plenty to say about Nothing—even the idea that there is a soul at all is examined. And just about the time the Greeks were laying their foundation stone, and just after the time of the Egyptians and Qabalists, the first soul-physicist was born. He was known as the Buddha—the Enlightened One. And what he said shocked us, and still shocks us today. There is no soul, he said. And yet . . .

CHAPTER 8

The Buddhist Nonsoul

A phantom, dew, a bubble
A dream, a flash of lightning, and a cloud.
Thus we should look upon all that was made.

—The Buddha[1]

PART 3 MIGHT SEEM TO contain much ado about nothing. Indeed, it has a great deal to say about illusion and reality. Here we will explore more deeply how it is that nothing can be the generator of something, and we shall postulate that this cannot occur unless there is also some form of intelligence associated with the process. Yet, to grasp this postulate, we will need to abandon nearly everything we thought we knew about the soul. The abandonment process has already started in part 2 as we began to see that in our soul-searching we were barking up the wrong tree, asking the wrong questions, and in doing so, ending up more separated from each other and from our souls than ever before.

At the end of part 2 we were left with our souls bared, stripped of all dignity, naked to the world of nothingness, mere spinning bits of virtual reality. When all else fails, it is back to square one. In our soul-deconstruction process we were able to see one clue as to how the soul accomplishes its reconstruction: the ability to separate, to demark the reality zone of spacetimematter from the twilight zone of nothing-but-potentially-able-to-be-spacetimematter that arose as a result of the uncertainty principle and the unusual game that knowledge plays with matter. That which is known secretly creates spatial separation from that which is known publicly.

By remembering itself, a character known as *the self* arises as if it pops out of the vacuum like a magician's rabbit from an empty hat. The *soul*-reality creates the *self*-illusion. The character takes on a life of its own. The story applies to every one of us. The writer of the play vanishes as his or her characters speak their lines pouring from the writer's fingertips. Everything seems okay if the writer and his characters remember their connection: who they really are.

We are like the characters in a play written by God. The danger in the confusion of the story of our lives is that we forget the soul. We tell and sell ourselves to the world. We separate self from soul. We become self-consciousness. We answer the wrong questions correctly and feel alone. To fall into space, time, and matter was hard enough, but now the soul believes it is the material self. In one of her songs, Madonna says she's just a material girl living in a material world. The soul is lost, and illusion reigns supreme as the characters continue to create *Much Ado About Nothing*, nearly totally forgetting that they were figments of the soul and writer, the Shakespeare who invented them all. It is as if Romeo and Juliet ask, "Who is William Shakespeare?"

So we need to return to nothing and learn how to generate something. We need to see literally how our souls are constructed: how and why they arise from nothing. We need to continue our inquiry into the soul-self dichotomy.

We are viewing a new picture of the soul, one that shows how it originates in empty space. This picture, a bather emerging from the sea, resembles the soul arising from the vacuum of space. And yet in this new vision an old question still reverberates in our minds. Could it be that the vision of a soul emerging from nothing is nothing more than a convenient illusion? Is the soul, like Old King Cole, then, also a mirage? Is the soul only a convenient label for the unimaginable processes of the emergence of something from nothing?

These and other musings about the illusory qualities of the soul lead to the concept of *anatta*, no-soul, as first explained by the Buddha. Surprisingly, they also lead to our modern physics account of the soul. Perhaps Siddhartha Gautama (later known as the Enlightened One or the Buddha) became the first soul-physicist when he denied its existence.[2] I use the term physicist to describe him because he questioned everything he could empirically test for himself and stopped asking about anything he couldn't. So Siddhartha began his inquiry into nature by denying anything assumed but not proven, such as the reality of the unproved soul.

Perchance I do the Buddha an injustice. Maybe he saw that although there is something real about the soul, human beings would never see it because they are attached to a particular picture of the self—a self that he saw, clearly although with difficulty, is an illusion. Perhaps he realized that the thinking of his time, dictated by the pundits and scholars who believed blindly in the soul, was clouded by the illusion of the self.

To grasp this subtle distinction, to see how the soul can be real while the self is an illusion—a combination of genuine self-illusion with imaginary soul-reality without making either into an oxymoron—requires some work, especially since in our soul-reconstruction program we need to pay attention to the essential difference between soul and self. That's what I meant by the above analogy of the writer and his characters.

The best way to grasp this new and original vision of the soul is to look at the major elements of Buddhist and ancient Chinese thought dealing with the issue of the illusion of self and compare them with my modern physics conception of the soul-self distinction. I hope to accomplish this here and in the next three chapters.

A Brief History of an Enlightened Soul Physicist

Looking back in time at the Buddha invariably leads to speculation, yet if his teaching means anything at all—and as we see later, it indeed has relevance for our lives—this instruction must be as important today as it was in his time, the sixth century B.C.E.

The story of the Buddha is of fairy-tale proportions. He was born a prince into a life of luxury until he suddenly questioned everything when, after leaving his luxurious palace life, he saw an old man for the first time. How can anyone grow old, become ill, and die? And why should there be a universe in which this happens?

Good questions! Many of us today are probably wondering the same thing. Wouldn't it be wonderful if life were a perpetual teenage romance or an adventure of youth?

Or would it? The continuing desire to move through life as if one were on a romantic holiday inevitably leads to unhappiness, as many of you who have grown wiser with age know. Ultimately, something happens which upsets all of our plans for blissful existence. We lose our jobs, fall out of love with our mates, have unexpected illnesses, see our loved ones die, and soon enough, become ill and die ourselves. If we haven't come to peace with this impermanent existence, we often die in regret. But before we do, we go on trying to see the upside of everything as usual, looking blissfully at the rainbows of life and ignoring the thunderclouds.

Seeing Through Illusion

Sad but true, you say? Sad but an illusion, I reply! Certainly life has its beauty and its magic, but life also has a tragic side, an edge most of us admit exists, even though we put our heads in the sand to avoid its falling on. We come to this edge in different ways. Some come in anger, some come in peace, and others come in sorrow. Yet, without our even knowing it, that edge, no matter how sharp it was, dulls in time. The edge of tragedy, like its comic opposite, seems a natural consequence of sentient life itself. Without comedy and tragedy, without laughter and tears, we would truly become unfeeling machines, and the whole notion of soul would become a pipe-dream fantasy with no hold over us. But we do feel for one another; we have compassion or come to learn compassion; we become aware of the common suffering of all sentient life-forms, including animals as well as ourselves.

Instead of asking obvious but wrong questions as to what causes this misfortune or that illness or that good luck or this horror, we come at last to the right questions through compassion and wisdom. We recognize that answering questions such as why something happened whenever a particular tragedy occurs rarely comforts anyone and continues to fuel our illusions and our deceptions. Ultimately through our suffering we are led to knowledge of the unity of the soul.

The First Spiritual Physicist

This knowledge of suffering and ultimate unity I call compassion. The Buddha saw it many years ago in spite of his wealth, pleasure, and position in society, and we must all see it today regardless of our positions in the world. The failure to see compassionately is a lack of thoughtfulness, a reversal in evolution, and we know from history, both personal and global, that we are capable of losing thoughtfulness as we move from birth to death.

Awareness of such a loss in mindfulness and a growing ennui with his uncompassionate existence wasn't enough for Siddhartha. He had to understand it in a deep way. We need to understand it even more deeply now.

We can view the Buddha as the first spiritual physicist! He worked toward the understanding of life and compassion in much the same way a physicist attempts to comprehend the world surrounding her. First he looked for fundamental principles by which he would be able to predict the behavior of the world around him. In other words, he sought the laws of existence, not just inert material existence but all existence, including sentient life. In so doing, he discovered a key insight: the universe is *dukkha*.

Buddhist Physics: All Is Dukkha!

Surprisingly, dukkha is rarely discussed in popular accounts of Buddhist thought, but I wish to explain this concept in some detail. Dukkha can be expressed in axioms that connect directly to our understanding of

Figure 8.1 Noble Truths: A Vision from Thailand. The Soul and the self, a perpetual dialogue ensues in quiet meditation.

the quantum-mechanical, classical-mechanical, and thermodynamical laws of the material universe, including its creation from nothing.

The word *dukkha,* interestingly enough, means a poorly made axle. Any wheel attached to it is quite likely to go awry, and thus dukkha implies things are not functioning as they could—there is a definite imbalance. According to the Buddha, the minute that life takes sentient form, all things go out of kilter.[3]

The Four Buddhist Axioms of Creation: Noble Truths

There are four basic axioms, called Noble Truths, which explain how things get out of balance. In actual fact there is really only one axiom: the reality of dukkha. The other three must connect to or be derived from it.[4] Let me out line all four Noble Truths:

1. There is dukkha.
2. Dukkha arises.
3. Dukkha ceases.
4. There is a path leading to the cessation of dukkha.

Most scholars refer to dukkha as suffering; however, suffering is a direct consequence of dukkha rather than an equivalence. Dukkha refers to the principle of imperfection, impermanence, emptiness, and insubstantiality. As a physicist, I would add to this list: uncertainty; indeterminism; constant change or flux; the arising of chaos from order and order from chaos; discontinuity; the vacuum state of all things; the quantum nature of reality; the principle of uncertainty; the probability nature of reality; physicist Ludwig Boltzmann's unit of entropy, k, which governs the relationship of the transformation of ordered energy to chaotic temperature; and the existence of physicist Max Planck's unit of action, h, which governs the knowledge uncertainty inherent in all material things.

In thinking about dukkha as an axle out of balance, one might be tempted to see these noble truths as indicating the commonality of suffering caused solely by physical imperfection. However, I believe the Noble Truths are actually related to how sentient life creates illusion. Certainly, all human beings are physically doomed. We all die for physical reasons. We all have our physical axles out of kilter, and the best we can do is endure the punishment. We are all born this way. We all begin the death process the instant we are created. Medically speaking, it is rather fantastic that we live at all.

Being out of whack, physically so to speak, has spiritual consequences. As the body deteriorates, the soul loses interest in it. One might think of this situation as the ultimate spiritual chiropractic position. The spiritual system will continue to go out of kilter as the body does unless one continues to go to his or her spiritual chiropractor. We need to "crack our spiritual backs" to let the light in once in a while.

One might get the impression from dukkha that all Buddhists are miserable and forever suffering. From my experience of them, nothing could be further from the truth. Indeed most practicing Buddhists that I know are quite happy, high-spirited, and live what appear to me to be productive lives. It is just that they seem to be unattached to anything and recognize or believe in the highly impermanent world as a way of life. They seem to enjoy both the happiness of sensual pleasure and the happiness of the absence of sensual pleasure, the happiness of pain and the happiness of the absence of pain.[5] The key to happiness is in allowing whatever arises to be as spontaneous and joyful as possible, recognizing that whatever occurs is impermanent and nothing, no matter how painful or pleasurable, is going to last.

Inertia, Entropy, and Action: The Physics of Emotion

To understand how dukkha works in our daily lives, we must see how it governs both the joys and the pains of life, the physics of our emotions. When we include concepts from physics this is quite illuminating. Dukkha causes:

1. Garden-variety suffering that we all experience when things remain as they are or change as predicted. *The first law of emotional inertia;*

2. Garden-variety anxiety and excitement we all feel eventually fade away. *The second law of emotional entropy;*

3. The action of events that moves us, which we experience as conditioned reality. *The third law of emotional action.*

Garden-variety suffering includes everything from having to smell the breath of someone with a bad case of gum disease to the loss of a loved one— all forms of physical and mental distress. We can refer to this as the first law of emotional inertia. As long as things remain the same or change as we expect them to, no matter how bad they are, it is better than having anything at all upset them. This is akin to Newton's first law of motion: bodies tend to remain in either a state of rest or of movement in a straight line.

Something apparently unexpected happens. We feel excitement or anxiety. But then our feelings change. Time passes and we find our joys or sorrows turning to blahs, our ah-has turning to ho-hums. Such is the dukkha of change. Thus the dukkhas of suffering and change are no mystery to most of us. This is akin to the second law of thermodynamics:[6] eventually entropy makes its way into all order creating chaos, and as it is in the physical world, so is it in the sentient world. The emotionally sharpened blade eventually dulls, the shiny metal eventually rusts, the ripe apple eventually rots, and the sharpest memory eventually fades. All things pass by, eventually becoming fuzzy.

The third law, the action leading to conditions of ever-changing physical and mental forces, is the most important. In it lies a clue to what the Buddha meant by the I, the soul, the ego, or the self and how he came to dismiss these concepts as pure illusion. Think of the I as mass. Consider the law as Newton's second law of motion: for a given force there is always a given change in the motion, provided there is a mass that experiences it.

For a given mass, a large force produces a larger acceleration or change than a smaller force. Big things affect us more than small things. We all know this. Major disasters, earthquakes and so forth, are certainly more destructive than the toilet's backing up.

For a given force, a large mass experiences a smaller acceleration than a small mass. Thus by buttressing up our egos, we suffer the fewest changes. By letting down our guards, getting that weight off our shoulders, we are able to change. Each day something in us, the I, is buffeted about by emotional forces. The effect it has on us depends on how much I we hold on to. The conditions we put before ourselves shape the I. These conditions limit our thinking: They are shields we put up to protect ourselves or shields we let down to free ourselves.

For the Buddha our big illusion arises in our attaching significance to the I. Even the thought that I exists is nothing but the arising of dukkha, and the concomitant result is emotion, in this case the emotion of the mind arising as thought.

Is You Is and *Is You Ain't My Baby!: Quantum Thinking*

Many believe that emotion tends to muddle thinking. In fact, just the opposite occurs; it tends to set thinking in motion by resolving the paradoxes of earthly existence. In what follows, I attempt to explain this and show how you can use what I call *quantum thinking* (QT) to avoid falling into the pattern-trap of fixed thoughts that occurs when paradox is resolved.

The usual difficulty we have with Buddhist logic and our misunderstanding of Buddhist thought arises because we normally use so-called Western or Boolean logic which incorporates only two values, right and wrong, yes and no, or zero and one.[7] Similar difficulties arise in our ordinary use of logical thinking about ourselves. In fact, that is precisely the problem. You can begin to answer the right questions when you possess the right tools with which to answer them. Western logic is not only not good enough to do this but also, unfortunately, is the problem itself!

For example, why do Buddhists not believe in the soul? That's simple, you might argue, they see all life as impermanent conditioned processes, and therefore, there is no need to have such a concept. However, there is more to this than meets the eye. For one may ask, why don't the Buddhists believe in the nonsoul either?

I can hear you: "What? What is a nonsoul? Is a nonsoul nonsense? Are you just playing with semantics here?" I assure you that I am not and that the idea of nonsoul, anatta, needs to be considered carefully in this new way of thinking.

To do this I want to apply quantum thinking to the old Newtonian and classical physical way of seeing. We could call the old way of thinking *Is you is* or *is you ain't my baby?* The answer to the question is allowed to be one or the other, but not both. In our new way of thinking we would ask instead, *Is you is* and *is you ain't my baby!* Here the answer is allowed to be both, and that leads to the resolution of all famous *koans* of Buddhist thought.[8] That is, we need to avoid defining things as having or not having a particular property by allowing them to have both. This is done by applying the principle of superposition from quantum physics to the game of consciousness.[9] This quantum thinking, when used in the game of consciousness, results in the avoidance of material addiction.

In the next chapter I explain in detail how material addiction can be seen in terms of Buddhist and quantum thinking. It appears that some *one* becomes an addict. But what becomes addicted? Is it just the body itself? Does the human body contain predilections for addictive behavior, a central addiction focus of consciousness perhaps? Would Buddhist doctrine state this and thereby admit the existence of something within the body like a soul? This might seem strange since according to Buddhist teachings there is no such thing as a soul.

What should you do to avoid such traps? Allow any feelings you have about any situation involving your self simply to fade away. Let any thoughts about the situation pass, as if you were watching them on a movie screen. Let the second law of emotional entropy play its role. Seems simple enough. What you are accomplishing by doing this is quantum-magical. You are suspending all action arising from defining the situation to run the show. You are maintaining the paradox and letting the projectionist in the projection-booth run the movie of your life.

On the other hand, when you feel anything arising and you allow action to take over, then eventual resolution takes place. Think of this as two cats at night attempting to establish territory. They howl and wail at each other, until one of them moves. Then the other gives chase, the wailing stops, and the situation is resolved. In a similar manner all of the egos within you wail when an emotional situation arises. The self/non-self illusion arises with the emotion. When the chase ensues, the soul falls deeper into matter.

What the Buddha Taught About Nonsouls

Let's return to the Buddha and review the concept of the soul enduring in his time. Next, we will look at the parable of the wanderer who had to deal with the inscrutable logic of the Buddha and then compare this logic to J. Robert Oppenheimer's aphorism on the equally inscrutable electron.

The term *soul* suggests that in each person there resides a permanent, everlasting, and absolute entity, an unchanging substance which lies behind every phenomenon in the "real" world. In Sanskrit this is called *Atman*. According to Hindu traditional discipline, Atman goes through many lives as it journeys through time, each time purifying itself until it finally is pure enough to unite with God or *Brahma*, taken to be the absolute Atman or Universal Soul from which it originally emanated. For other spiritual disciplines, particularly Western religions, the soul lives eternally in heaven or hell, depending on the whim of God, who created it in the first place, journeys to the earth for a single lifetime, and then returns to either place for eternity.

Western religions rarely deal with reincarnation, although they do deal with resurrection, but Hinduism has no problem with it. You would also think Buddhists would not cotton to reincarnation. But in spite of their belief that the soul does not exist, the Buddhists do believe in reincarnation. How can there be reincarnation if there is no soul that reincarnates? To grasp this we really need to use QT, which allows us to hold two sides of a paradox simultaneously. In chapter 11, I'll explain reincarnation using QT. Meanwhile, let me continue with my story about the paradox of the nonsoul as a lesson in QT. This paradox is illustrated in the story of the wanderer.[10]

One day a wanderer came into the village where the Buddha taught. His name was Vacchagotta. He asked the Enlightened One whether or not there was a soul (Atman). The following was their somewhat brief and one-sided conversation:

VACCHAGOTTA: Venerable Guatama, is there a soul?

BUDDHA: (Silence.)

VACCHAGOTTA: Then Venerable One, is there no soul?

BUDDHA: (Silence.)

VACCHAGOTTA: (Gets up and goes away.)

Later Ananda, a disciple of the Buddha, appeared and asked the Enlightened One to comment on his previous silence. The Buddha said:

Ananda, when asked by Vacchagotta the Wanderer, "Is there a soul?" if I had answered, "There is a soul," then that would be siding with those recluses and brahmanas who hold to the eternalist theory.[11]

And when asked by the wanderer, "Is there no soul?" if I had answered, "There is no soul," then that would be siding with those recluses and brahmanas who hold to the annihilationist theory.[12]

Again, Ananda, when asked by Vacchagotta, "Is there a soul?" if I had answered, "There is," would that be in accordance with my knowl-

edge that all dhammas [ways of inquiry, paths to enlightenment] are
without soul?[13]

And when asked by the Wanderer, "Is there no soul?" if I had answered, "There is no soul," then that would have been a greater confusion to the already confused Vacchagotta [who earlier had inquired
into what happens after death and was confused by the Buddha's an-
swer]. For he would have thought: "Formerly indeed I had a soul, but
now I haven't got one."

We can compare this legend with one well known from quantum
physics. One day a student wandered into the chambers of Robert Oppenheimer, the physicist who in the 1940s headed the scientific team that constructed the atomic bomb. As the story goes, the student asked Oppenheimer
about the existence and movement of the tiny subatomic electron within the
confines of the atom, to which Oppenheimer answered:

> If we ask, for instance, whether the position of the electron remains
> the same, we must say "no."
> If we ask whether the electron's position changes with time, we must
> say "no."
> If we ask whether it is in motion, we must say "no."
> If we ask whether it is standing still, we must say "no."[14]

Oppenheimer's quote and the Buddha's response to Ananda regarding
the soul point to the same thing. For in both Buddhist logic and QT, it is necessary not to hold any fixed opinion but to see things as they are without
mental projections. Both Oppenheimer and the Buddha refer to the condi-
tionality of all things, whether they are soul essence or electrons. The Buddha suggested what we call I is nothing more than an aggregate of processes
working together in a flux of momentary change within the law of cause and
effect, and nothing is unchanging and eternal in all of human experience.

Oppenheimer also reminds us that *electron* is nothing but a term of convenience, a label that points to something we infer exists but which defies the
usual explanation. It defies it because it is not a real thing and has no real substance.

The Tibetan Buddhist Soul Returns

The question of the Wanderer still haunts us. Without a soul, what happens to
me after I die? Perhaps the answer is hidden in modern Tibetan Buddhist practice and can be found in what the Dalai Lama does. This ancient tradition,

which the Communist Chinese are trying to eradicate in Tibet, is based on an individual's training his or her mind to understand the deepest levels of existence.

The Dalai Lama, in describing these levels, states:

> My daily practice is preparation for death through meditation—to make a separation of the body and consciousness. Unless you reach the deepest sub-consciousness, you cannot separate this body and mind. When you reach this deepest state, then they can separate. Then I go deeper, deeper, deeper to the deepest which is the clear light.
>
> Sometimes I joke: In 24 hours I experience death and rebirth seven times. Then when the actual death happens, this practice becomes very useful.[15]

And, we archly ask, useful for whom, a soul?

What the Buddha Said About the Soul

Scientists are not the only ones who have trouble with consciousness and the existence of the soul. The Buddha once chastised one of his disciples, Sati, for misquoting his teacher by stating that consciousness transmigrates and wanders about. The Buddha stated:

> You stupid one, who have you ever heard me tell that nonsense to? Haven't I always told you that consciousness arose from conditions, and that without these conditions there is no consciousness?[16]

Next the Buddha, perhaps somewhat impatiently, laid his cards on the table and explained how one apparently can have a soul without having one. Better than any scientist, he demonstrated consciousness to his student, and from this definition we, as scientists, could begin to look earnestly. He said:[17]

> Consciousness always carries a label. In and of itself, it has no object. It is named according to how it arises. Because we have eyes, we see visual objects and we have visual consciousness; because we have ears, we hear audible sounds and we have aural consciousnes; because we have noses we smell aromatic objects and we have olfactory consciousness; because we have bodies, we feel touch-sensations and we have tactile consciousness; because we have tongues, we taste foods and the like and we have gustatory consciousness; and because we have minds, we think mind-objects (ideas and thoughts) and we have mental consciousness.[18]

The Buddha was describing a being or entity. He was, as we are, concerned with the definition of the self, soul, individual, subject, or simply the I. To him, and to a great extent, to many scientists, this beingness is not something permanent. In itself, it has no real substance but appears to arise as a combination of ever-changing forces of transforming mental and physical energies (the third law of emotional action).

These energies are grouped into five very specific aggregates called *pancakkhandha*, and they all have to do with attachment, our predilection to have and hold on to things even if those things are imaginary. In other words, without attachment, there is no being and beingness vanishes whenever attachment disappears. (In the next chapter we see how the laws of attachment work in creating something from nothing, and in chapter 11 I tell you more about how these aggregates combine to produce reincarnation of a soul!)

Taken together as arising and vanishing each moment, these five aggregates constitute the grounds of being that we attribute to our soul or our self, and here no distinction between "self," "soul," and "ego" is made. They are synonymous in Buddhist thought.[19] Each aggregate can be imagined as an interlinking mechanical part, ever turning and interacting and in its motion ever causing the others to turn. The idea of I is thus a false appendage or, at best, an epiphenomenon arising from the interdependence of the aggregates.

Zen Mind, Zen Body, More Quantum Thinking

Some of the wonderful aspects of Buddhism which I truly delight in are the paradoxes of Buddhist thought. Nothing is more paradoxical than the form of Buddhism known as Zen. According to Buddhists,[20] Zen Buddhism arose as a Chinese-Japanese branch of Mahayana Buddhism and is based upon the teachings of Bodhidharma, a far-wandering iconoclastic and apocryphal Indian Buddhist master who lived more than one thousand years after the Buddha. Apparently a Chinese emperor named Wu demanded that Bodhidharma appear before the Nanking court. If we imagine we are in a time machine and go back in time to around 500 C.E., we might overhear this conversation between Wu and Bodhidharma upon the latter's arrival at court. We should remember that by this time in history, Buddhism was well known in China and the emperor, once having been an ardent Confucianist, was now an ardent Buddhist. Here is their conversation (again my reconstruction):[21]

EMPEROR WU: Welcome to the Great Court of Nanking. I have long awaited your arrival. As you see I have done well by Buddhism. No animals in the kingdom are killed, capital punishment is coming to an end here, and many Sanskrit manuscripts have been translated into Chinese to "spread the Dharma!" Even I read to my court members from the Buddhist scriptures. I have also built many Buddhist temples in the land.

What merit have I accrued? How much bad Karma have I wiped out?

BODHIDHARMA: None.

A hush fills the court. Bodhidharma has spoken very abruptly. He is not polite to Emperor Wu. The emperor looks concerned. He paces the floor. Anger seems to be welling up in him. He turns to Bodhidharma and speaks.

EMPEROR WU: None? How can that be?

BODHIDHARMA: No merit whatsoever.

EMPEROR WU: What among the holy teachings of Buddhism is the most important then? What is the First Principle of Buddhism?

BODHIDHARMA: Vast Emptiness and nothing holy.

The emperor is astonished. It is clear that he is nettled and is holding in his anger. He looks at Bodhidharma with a mixture of curiosity, hurt, and superiority in his eyes.

EMPEROR WU: Who are you who replies to me in this manner?

BODHIDHARMA: I do not know.

At this point Bodhidharma leaves the court.

According to legend, Bodhidharma crossed the wide Yangtze River to a distant cave where he sat for nine years. And this is how the Zen sitting meditation called *zazen* arose. Bodhidharma's truncated responses may of course have been heightened by history. He had little regard for long involved discussion. In denying merit-accrual, Bodhidharma was rejecting reward for good works and stating that overattention to such forms as rituals, temples, good thoughts, and monuments would never bring liberation from suffering. To liberate, sit.

To Be Liberated, Sit and Enjoy the Void

The vast Emptiness stressed the essential idea of *void*: nondualistic and eternal. This void can be called the *eternal self* or, as I shall describe it later, the *unreflected spirit*. It would correspond to the Buddha's idea of reality as distinct from existence. In Zen training in its earliest stages aspirants are told to "empty the mind—take as thought, the thought, 'no-thought.'"

When asked about his identity, Bodhidharma said there is no-self here at the precise instant of spontaneous existence. Thought arising creates the illusion of anyone's being home. This paradoxical way of thinking is popular in Zen today and appears in the form of koans. For example:

> To study the Buddha way
> is to study the self.
> To study the self
> is to forget the self.
> To forget the self
> is to be enlightened by all things.[22]

To which I would add:

> To be enlightened by all things
> is to study the Buddha way.

This almost completes the round of enlightenment. An important nuance was added to Buddhism by Bodhidharma, the concept of not asking stupid questions. A stupid question is one that by its nature cannot be answered. There is instead a renewed stance toward trusting one's own experience. If you can't answer the question for yourself, it is a stupid question. Bodhidharma could not say who he was. Why was that? Because there was nothing within him that he could identify with. That is why no-self is the answer. No one home.

Bodhidharma also said:

> No dependence on words and letters.
> Seeing directly into the mind of man,
> Realizing true nature, becoming Buddha.[23]

"Just doing it" is the quality of Zen. There is no mind, no body to distinguish. "The way is attained through the body" because of "the unity of mind and body," said the thirteenth-century Zen master, Dogen.

These koans delight the mind because of the seeming nonsense contained in them. They are, if you will, the stupid or wrong questions I mentioned before. The example, probably well-worn by now, "What is the sound of one hand clapping?" is ridiculous. Yet we seem to know what it means.

When asked this one day by students in a class, I spontaneously began applauding. The students said I cheated because I used both hands. I responded by asking, "Doesn't one hand clasp the other when I hold my hands together?" The students responded affirmatively. "Then," I asked, "doesn't it follow that one hand claps the other just as one hand clasps the other?" Again they said yes. "Well then that's the sound of one hand clapping. It's the sound made when it claps the other." Of course, the students immediately objected, and I don't blame them.

It may be hard for our Newtonian, mechanically trained Western minds to appreciate, but Buddhists claim all words and explanations are like my "clapped hands" example. Every answer leads to another objection in turn.

> Not equal to
> Not metaphor
> Not standing for
> Not sign.[24]

The idea in Zen is direct unity, no substitutes. Maezumi Roshi, who heads the Zen Center at Los Angeles, wrote:

> What the Buddha had experienced was . . . the direct and conscious realization of the oneness of the whole universe and of his own unity with all things . . . To have this very realization is in itself to be the Buddha.[25]

Existence *is* because of the process of *reflection*[26] arising from reality. The Buddha, in contemplating existence and reality, was wrestling with his koan.

The *Not equal to* koan was like the Buddha's meditation under the Bodhi tree. He suddenly realized the meaning of unity in all things. He became the grass, the tree, the setting stars. All questions stopped. He saw reality. He saw beyond all distinctions. He did this by just sitting.

Shunryu Suzuki Roshi said:

> When we have our body and mind in order, everything else will exist in the right place, in the right way.[27]

He also wrote:

> What we call "I" is just a swinging door which moves when we inhale and when we exhale. . . .

When your mind is pure and calm enough to follow this movement, there is nothing: No "I," no world, no mind nor body; just a swinging door.[28]

Zen means unambivalent concentration or absorption. The lack of ambivalence is important. It means the avoidance of all attempts to make distinctions, to be practiced with all beingness, all mindfulness, and with no distinction between beingness and mindfulness.

The Humor of Nonsouls

The most endearing quality of Zen is its sense of humor. One story has it that the third Patriarch died while standing upright with clasped hands, not a normal posture for death at all. Another aged sage asked his disciples, "Who dies while sitting?" The students answered, "Enlightened monks." To which the old one rose and walked, asking the monks to question the validity of such stereotypes. On the seventh step, he died with his hands dangling at his sides.

Another monk named Ten Yin-feng was determined to die in some outstanding fashion. Just before he passed he inquired of his disciples if anyone had ever died standing up. Being assured that this had taken place, he asked "How about upside down?" No one had ever heard of such a demise. So Ten stood on his head and died. Crowds of people came to see the dead master standing on his head. This was a great inconvenience to all of his fellow monks who could not agree what to do with Ten's remains. At last Ten's sister, a Buddhist nun, arrived on the scene. Viewing her brother's corpse as it presently stood, she said, annoyed, "While you were alive you took no proper notice of laws and customs and even now that you are dead you're still making a nuisance of yourself." Whereupon she gave Ten's corpse a shove, knocking the body over so that it could be carted off to a proper burial by his brothers.[29]

A Brief Look Ahead

Now that we have had a look at anatta, it is time to see how manifestation of anything must lead to good and evil, pleasure and pain, addiction and recovery, sorrow and happiness, and matter and antimatter. For in the universe we must have an essential duality as the Spirit of nothing attempts to reconcile with that which is something. The battle of time and timelessness never ceases. The vacuum never ends as it attempts to recover the matter that spewed forth from its guts just a few billion years ago.

Richard Feynman, the famous Nobel Prize–winning physicist, noted that particles that live for extremely short lifetimes are called *virtual* particles because of their early demise. But how short is short? From one point of view, even so-called real particles are just longer-lived virtual particles that will see the end of their lives when the big crunch comes.

Until that time we need to put up with the vacuum's attempt to "straighten us out." But in doing so we are faced with alternatives, choices, from which we can't assume any outcome. This gives rise to the essential duality of life we all feel. This is the ultimate "reason" behind the greatest good and the basest evil. In the next chapter we look at Good and Evil, the temptations of the soul (even if *it* doesn't exist itself).

CHAPTER 9

Good, Evil, and Soul Addiction

You like my puppets? If you want to understand Java, you have to understand the Wayang: The sacred shadow play. The puppet master is priest. That's why they call Sukarno the great puppet master, balancing the left with the right. Their shadows are souls and the screen is heaven. You must watch their shadows, not the puppets: The right in constant struggle with the left, the forces of light and darkness in endless balance. In the West we want answers for everything. Everything is right or wrong, good or bad. But in the Wayang no such final conclusions exist. Look at prince Arjuna. He is a hero, but he also can be fickle and selfish. Krishna says to him, "All is clouded by desire, Arjuna—as a fire by smoke, as a mirror by dust . . . It blinds the soul."

—Billy Kwan in *The Year of Living Dangerously*

IN CHAPTER 5, following Dr. Frank Tipler's vision of the computer soul, we saw that heaven, the ultimate abode of goodness, and hell, the destination of evil, will both arise at the end of time when we will have time to reconsider what life, the universe, and everything is all about. Maybe then we will have figured out why good and evil exist.

In the previous chapter we took a good look at how Buddhists and quantum physicists may come to some agreement about the absence of any soul and how everything spontaneously arises from dukkha. In the attempt to re-balance the unbalanced axle that dukkha refers to, we seem to make matter(s) worse. In our attempts to do good and right the world's wrongs, quite often we find ourselves doing harm. For example, our use of pesticides, while solving a hunger problem, created a health hazard. Going to war for the noblest purposes usually creates future problems for both sides, regardless who wins.

Up jumps the devil taunting us as soon as we feel satisfaction from our good deeds. Out of everything that occurs appear the clear signs of good and evil. They seem to be as necessary to each other as the head and tail of a coin. The hard thing is to recognize when we move from the light into the shadow, for the dark side is not always clearly visible.

Do good and evil really exist? Or are these, as well as the self, illusions? According to Buddhist thought, there is no evil, only misguided thinking which leads to evil deeds through ignorance. What about good? Equally illusory say the Buddhists. Out of ignorance good and evil arise as a single unity. What causes such ignorance? I wish to address the key concept in chapter 9 of ignorance springing from *addiction to illusion.* Here we will look at the nature of good and evil as arising from such an addiction, a process that I suggested in chapter 8 originates from dukkha.

Dukkha produces suffering. Suffering inevitably produces illusion. Any evil or good is ultimately an illusion arising from this suffering. When dukkha ceases, suffering vanishes. When suffering ceases, illusion fades, good and evil vanish. The appearance of good and evil is an illusion caused by misunderstanding the suffering that one feels.

For example, one murders another. The illusion would be that the other had something that the one wanted. We can imagine several possibilities: The murderer suffers from the illusion of helplessness, an inability to sustain his life by doing service, necessary work. He suffers from the illusion of low self-esteem and from the illusion that he is alone in the world, cut off from all humanity. He believes he is a unworthy, perhaps even less than human. He does not see that taking a life *kills him* by causing needless mental anguish. He does not believe this because he thinks he is alone and his mind is unique, separated by skin and bone from all others. As I shall demonstrate in chapter 12, this is perhaps the worse illusion of all, one that many of us suffer from even though most of us do not murder on account of it.

As a result of the murderer's addictive thought patterns of unworthiness and helplessness he becomes addicted to his murdering actions and falls deeper in despair. The murderer fails to see suffering as it is. Instead he has used his suffering as an excuse for murder. Paradoxically, he may even hold himself in high esteem as a result of illusory addictive images of himself. (I will have more to say about the illusion of self-esteem later.)

There is the other side of suffering's power to create illusion. This is rarely dealt with: the illusion of good, the opposite side of the coin. As with evil, suffering creates evil's opposite. Instead of the suffering of the murderer we have the suffering of the law enforcement officer. Clearly, he does good by catching the murderer. However, just as the murderer identifies himself as one who takes the life of another, the law enforcement officer must identify himself as the one who takes the life of the murderer. The dance goes on. To protect those of us who "feel" innocent, we have created an illusion called government that protects us from evil. The law officer serves the judicial system of that government, where, as current trends seem to be taking Western thought, the murder of murderers has become law in nearly every state of the United States. We call it execution. We call it good. We call it revenge. We live in fear of our neighbor, failing to see him as ourselves. We are addicted to this way of thinking and most of us see no way out.

Is there a physics to this addiction which creates the good/evil duality? Could it be that heaven and hell spontaneously arise from the vacuum of space, as naturally as the axle of dukkha twirls out of kilter and the object in the sunlight casts its shadow? If so, then we face our devils no matter how hard we try to escape them. But why is the universe so cruel? Heaven, okay, but why hell? What is hell anyway?

Two Forces in the Universe

Two opposing forces in the universe appear to anyone who has thought about it at all. We might call them the forces of light and dark, good and evil, or order and chaos. Everywhere we look we see these forces at play, acting on us and others, driving us to consider the effects of our actions in the world. We see these forces displayed in the physical world of matter and energy, and we attach moral or spiritual significance to the results of the actions of these forces. We express these actions in metaphorical terms of mass and inertia, sometimes, but not always, associating good and order with action and evil and chaos with resistance to action or passivity.

These two forces create nearly every duality we face. We are all continually pulled in opposite directions by countercurrents. We recognize that our souls are at the centers of these maelstroms where we find ourselves in a battle, and yet we see this essential and personal center very dimly. We sense that it is present and we give it a name. We call it *soul* or essential *self*, and we recognize that it may indeed suffer from desire as we go within and confront our deepest longings.

Who has seen a courtroom drama and not remembered the parting remarks of the judge to the prisoner convicted to death, "May the Lord have mercy on your soul"? We believe ourselves innocent as babes, in terms of crimes as inhuman as murder, and wonder, how could anyone kill? How could anyone consciously maim or injure another human being? We behave as if these actions are beyond us. We remember Poe's *The Tell-Tale Heart* or Dostoyevsky's *Crime and Punishment*, fascinating ourselves with the dark side of the suffering soul. Yet in spite of our seeming innocence, we find evil in all its facets fascinating and sometimes compelling.

Who Knows What Evil Lurks in the Hearts of Men?

Whenever we face compelling desires, we confront what Carl Jung called the shadow, and just as in the old-time radio show, we hear the shadow asking the above chilling question. Jung refers to the *shadow* as the inferior being within ourselves, the one who wants to do all those things that we would never do ourselves. Every culture deals with the shadow as the center of evil action. Frieda Fordham, a noted Jungian authority, puts it thus:

> The shadow is . . . a collective phenomenon [and] expressed as a devil, a witch, or something similar.[1]

We saw in chapter 4 that the shadow was of great concern to the ancient Egyptians. Their Book of the Dead warns that whoever holds the shadow of a soul can work evil against the soul. But what about us today? Jung made us face our shadows, indicating that if we didn't, they would creep up around us and even destroy us. Can we deal with the force of evil through our will and intent, or are we doomed to evil actions in spite of our wishes?

In the film *The Year of Living Dangerously*, as we watch ancient Javanese shadow puppets in the film's opening scenes, we realize each of us is no more than a marionette, following perhaps invisible strings but, nevertheless, manipulated by forces of which we have no apparent awareness or control.

We name these controlling forces *desire* or *compulsion*. When desires or compulsions overcome us, we feel out of control, and when they continually overwhelm us, we are faced with behavior we label as *addictive*. At these moments we believe someone else is pulling the strings; a force is acting on us that we are unable to manage. Indeed the famous Twelve-Step Program of Alcoholics Anonymous is based on the recognition that a single individual has no control over his or her addictive destiny.

Could the soul's suffering be caused by addiction? Can we relate good and evil to addiction? The answer seems to be *yes* to both questions. Here is how it all works . . .

The Soul and Its Shadow: Addiction

To exist in empty space as nothing but potential energy is not easy. *Nothing* is unstable! (Read the preceding sentence as if *nothing* were a thing.) This tendency to come into being is difficult to describe. Imagine a large football-shaped boulder balanced on one end while sitting atop a sharply peaked mountain. Imagine a turtle standing on the top end of that rock. If the turtle moves forward one inch, the rock will tumble down. Atop the turtle sits a flea. If that flea moves one millimeter, the turtle will be forced to lose its balance and the rock will come tumbling down.

The soul resides in the vacuum of space as the flea on the turtle. Its desire to manifest is its addiction. Resisting the temptation to manifest is like keeping the flea from moving one millimeter.

You might think of the soul's shadow as its addiction to matter, to coming into being. The soul is addicted by desire for its material image, much as Narcissus was so enraptured with his reflection in the river that he was frozen in place, and the dog with a bone in its mouth, who was so desirous of the image of the bone it saw reflected in a pool of water that it dropped its mouthful to fetch the illusionary bone. In what follows I'll show how such addiction—the desire to have an imagined thing—produces substance abuse, addictive behavior, leading to obvious evil, and, as mentioned above, the physical universe.

What is addiction? How does it arise? What becomes addicted? Is it the body, as we are led to believe by current medical models? Some models of addiction, based somewhat on biology, suggest there are two types of human beings, addicts and nonaddicts, and that addicts, victims of their own bodies, are truly powerless to prevent their addictive behavior on their own. How does addiction play a role in our life patterns? In this chapter, we examine some new and perhaps startling ideas about addiction and its relationship with the soul. These ideas are based on quantum physics, my experiences with shamans in the Peruvian jungle, and some ancient Buddhist teaching.

Soul Ecology and Soul Pollution

The soul can become polluted when it acts under the illusion of doing good or evil. One might be tempted to regard addiction as evil in itself. However, there is some deeper cause of soul addiction, something perhaps more insidious and yet more enlightening. Addiction is neither good nor evil, it is a process. Not recognizing addiction as a process, under the illusion that one is doing good by denying the addiction, often leads to despair when one sees the truth of one's actions. Those who have suffered the pains of addiction and have recognized it as a process may, indeed, be more aware of their individual souls than those of us who have not. Certainly, anyone who has recovered from substance addiction knows what evil arises in addiction denial and also knows his or her soul when that denial is finally dealt with.

Can we put the vulnerable soul at the center of good and evil, order and chaos? Why is the soul capable of pollution and, thereby, unrestricted deeds of evil? Why would a universe or a God in a universe place souls in this kind of jeopardy, at such risk, that concomitant addictive behavior resulting in evil action could and does occur? Can we come to some answer based on our present quantum view of the universe, matter, and energy?

I think so. And even though I shall refer to an ancient and apparently soulless way of looking at the human spirit to find this answer, it will shed light on the physics of the soul and its existence. What emerges from this study is, perhaps, surprising. The vacuum, under certain circumstances, has a greater tendency to spew forth energy and, ultimately, matter, than not to do so. This *vacuum compulsion* is perhaps the addiction to having something rather than nothing. Thus, here we shall explore the notions of a compulsive vacuum, addiction, and how together they cause human suffering.

The Shadow of My Soul: Addictive Desire

Addiction is a cosmic process, a reflection of the urge or Demiurge to create matter out of nothingness. This new vision of an antiquated problem traces its roots to a very old concept once brought forward by the Buddha. We may call it *false identity*. It is a process ultimately linked to the soul's material addiction and to the action by which the soul becomes clouded by desire—a necessary link in the Buddhist's concept of Conditioned Genesis. In brief, it is the vacuum's desire to form the I.

The vacuum creates virtual processes, illusions of material activity, all occurring in the negative energy sea of potential matter. It does this spontaneously, at times erupting into the presence of real matter and energy. However, this *real* matter is constantly bombarded by the virtual matter left behind in the negative energy sea.[2] The battle of matter and nothingness results in longing and desire to return to nothingness along with desire to erupt spontaneously into something once again. The battle continues and is felt as suffering, an axle out of kilter. The battle occurs on a time and space scale far beyond anything we have ever seen or can even imagine.[3]

Using the concept of the soul as that which suffers when it becomes the body, and following the paradoxical nonsoul teaching of the Buddha, I believe I can show that all suffering arises from the soul's addiction to substance, which in quantum physical terms means the vacuum's desire to have matter present—which arises from the vacuum's tendency to create light and energy, forming the basis of materiality. This addiction is the root cause of all suffering. I see this addiction as the fundamental mechanism by which spirit becomes matter.

Another way to put this is that spirit exists in the pure vacuum state. Buddhists call this state the void. Qabalists call it aleph. As we saw in chapter 6, this vacuum state continually froths up matter and antimatter, momentarily springing forth only to be gobbled up again.[4] What we call matter in this universe is the remnant of that bubbling, a gigantic bursting out called the big bang that supposedly occurred 10 to 20 billion years ago. Nothing created something then, and this process will continue forever so long as there is the potential to create something. That potential continues forever so long as nothing exists! Therein lies the secret of the spirit's longing: the desire to become something rather than nothing.

Spirit Trapping

Stable forms of matter and spiritual awareness are the desired outcomes of spirit. In order for matter to arise and spirit to become aware, spirit must create structures within itself. We might say that in the vacuum state, spirit dreams. Or we might say spirit creates a holographic illusion. Spirit's dynamic structure consists of levels of undulating potentially conscious and potentially unconscious movements, much as an ocean consists of wavelike movements on the surface and deep below. Through reflection of this movement, resulting in bubbles of matter and waves of unconsciousness and consciousness, stable repetitive patterns arise and spirit's desire for an experience of repeatable security (an impossible dream) continues and in doing so creates addiction. Spirit falls in love with its own reflection. Once started, universes pop into existence, big bangs occur, and omega points abound. Some universes last only a second. Some last for 20 billion, billion years.

When a universe pops into existence, it has immediately set in motion its own demise. For every starting point there must be a finishing point. For every big bang there must be a big crunch.[5] Today physicists aren't sure that our universe will come to a big crunch end. It is a question of delicate balance between matter and energy. From a spiritual point of view, it makes sense to have the universe come to an end.

Between the beginning of a universe and its end, time and space come into existence. Here the spirit, in its attempt to form stable images, begins to form waves. These waves, like ocean waves reflecting from a pier, bounce back from the beginning and ending points of the universe. This reflection becomes conscious as *soul.*

Remember the image of the turtle resting upon the boulder at the mountaintop. Things are delicately balanced. Any unfocused desire is likely to send the whole thing tumbling down. Creating stable images requires concentration, real effort if I can imagine spirit able to make such effort. With effort and concentration, a stable pattern forms in the void and repeats itself. The goal of the spirit is to repeat a vibrational pattern indefinitely and put it into space and time. A shortened persistence of the pattern results in dim awareness and brief flashes of matter, such as seen in high-energy particle events. A longer persistence results in strong awareness and stable particles of matter.

Whenever the pattern invokes longtime matter, creating atoms and molecules, an effort is involved. In a similar manner, secure memory arises in the brain through intent: longtime repetition of patterns. The actor remembers his lines with such an effort. The vacuum remembers her particles in the same way. For this security to occur, for a stable pattern to remain, an effort must be made to hold the image much as does a meditator, who, after watching a flaming candle in a darkened room, closes his or her eyes and holds the image of the flame.

Spirit trapping is the containment of spirit within spatial and temporal boundaries. Spirit trapped between the beginning and ending of time (big bang and big crunch) produces the soul in the same manner as a harmonic wave pattern is produced on a vibrating string held between two fixed ends. The same trapping within the bounded, but infinite, space of the universe creates matter and the world we all see and feel about us.[6] But the price for this, paradoxically and amusingly, is the simultaneous creation of uncertainty.

Patterns of Uncertainty

Here I make a major premise: Awareness literally requires reflection. Think of this as mirror reflection or the reflection of an ocean wave bouncing from a pier. All patterns require reflections. All reflections require boundaries. Hence, consciousness requires boundaries and its purpose is to create novel patterns.

Spirit becomes trapped within the boundaries of space and time in order to become aware of itself. However, these boundaries, temporal or spatial, are unstable. They waver like images of the candle flame in the mind's eye. When the boundaries change, the pattern of reflection changes. Just as a violinist creates different notes by changing the positions on the fingerboard—trapping the notes between the fingers and the fixed ends of the strings—so are new patterns of consciousness created by changing the boundaries trapping the spirit. As boundaries waver and dissolve, spirit emanates, bursts forth, and returns to the vacuum and becomes unconscious. The process repeats. This trapping and freeing of spirit results in the chaotic dance of consciousness, unconsciousness, matter, and energy called quantum physics. Because of the inherent material instability, nothing can be known with absolute certainty. Things just disappear and reappear on the scale of subatomic matter. So does awareness.

The uncertainty principle expresses in mathematical terms how embedded spirit appears with inherent probability in matter. When embedded in matter, conscious spirit yearns for freedom but has forgotten where it came from. Being spatially trapped, it can only "reflect" on itself within its spatial boundaries. It loses sight of itself as unbounded spirit.

Two forms of spirit trapping occur: temporal and spatial. The temporal traps at the beginning and ending of time reflect spirit's vibrations. The pattern of those reflections is the conscious soul.[7] The soul, as temporarily trapped spirit, is capable of the clearest awareness and the least illusion. When spirit is trapped in space, it forms matter. The soul can also "see itself" in matter as the reflection of spirit in space. When the soul desires this, the soul begins to lose sight of its own nature and begins its fall into matter.

Spatial and temporal boundaries forming in the vacuum continually trap spirit. Spiritual release from spatial traps allows the soul freer reflection: the soul becomes more aware. Release of spirit from its temporal traps results in the spirit's falling asleep: the spirit returns to nothing. To awaken, another reflection must occur. In terms of the Qabala, aleph, potentially able to be anything, becomes *yod*, existence, when spirit becomes temporally trapped. The war with time ensues. The spirit reflects on its new boundaries and becomes aware. When the time-trapped spirit, the soul, reflects itself in space, matter appears.

The soul is in constant danger of becoming a matter junkie. It can become addicted to its image as substance. The soul, as spirit, will continue its battle with its reflection in matter and attempt to free its image from material boundaries. This "image" is more than it might seem: It holds the soul, embedding it in matter. Ultimately, the soul will free itself, but just as quickly, like the unstable rock on the mountain, it will "fall" into matter again. As a conscious reflection of soul within matter it may not remember itself as spirit, and the battle of the soul with its spatial reflection, known as the war with time, continues.

Think of it this way: The vacuum is a bubbling cauldron of expectations, hopes, and dreams—of everything. Every once in a while the dream pops out of the vacuum in the form of particles of matter, and the vacuum attempts to retrieve it as soon as it can. But, like a bird freed from a cage, the created matter feels its wings and says, "no thanks." It cries out in illusion, "I'm free at last." That is how the big bang got started banging. Meanwhile, some 10 to 20 billion years later, we, in our spiritual quests, yearn to free our spirit and its bounded emanation called soul from the traps of such material addiction. We desire substance in our material quests—as much as we can get. The battle of time and timelessness never stops.

Addiction is a cosmic process, a reflection (literally) of something vastly more subtle and pervasive: the creation of matter and consciousness out of nothing, but the potential to have such things. Strangely, the theories of quantum physics, ancient Buddhist teachings, and visions of shamans in the jungles of Peru agree on this point. Addiction is universal to all life forms, and any addictive behavior arises as a result of the soul's desire for material form. Consequently, addiction cannot be cured without considering the spiritual dimension of human beings. We need to grasp that desire for a material form is the root cause of any and all addictions. Once the soul falls into matter, it becomes addicted, and it ultimately suffers seeing itself as the self. To relieve the

Figure 9.1. The Dynamic Cauldron of Spirit. Ever attempting to structure itself thus forming matter and soul, spirit paradoxically tries to free itself from material addiction; this dance of the spirit, like the proverbial Timex watch, keeps on ticking.

soul's suffering, spirit must return to its nonmaterial state. In a way, the trapped soul must give up its most precious feeling, that it is real.

We need to examine how this fall into matter occurs and why we suffer. The fall is necessary for a universe to arise and our sentient suffering ensues as night follows day.

The Fall of the Spirit into Matter:
The Soul Believes It Is Ego

The first sin, so to speak, is false identity, believing that we are matter and, in doing so, suffering. The Buddha felt all suffering had its roots in this one cause: the false identity of the multitude of processes within the body and mind with the ego or the self. Suffering is what is experienced as a result of investment within a particular ego's self-image. I label this a false identification of soul with ego or, in terms of a geometrical model, the "belief" of the volume of an enclosed space that it is its surface.[8]

What is a false identity of the soul with the ego? I thought I had enough problems dealing with a true identity, and now I have to deal with a false one? Let me explain with a story. It illustrates how a false picture can lead to suffering.

Once upon a time there were two monks. They had just spent several months in retreat where they had vowed silence, chastity, and renunciation of impure thoughts, and they now were making an arduous journey. One of them was quite young and the other was quite old. They came to a river, and as they were lifting their cassocks in preparation to cross the river, a sexy young woman appeared next to them at the shore, seeming upset. The old monk, breaking his vowed silence, asked her what the problem was. She said she had to cross the river, but she was afraid of water. The old monk said not to worry, he would be very happy to carry her in his arms across the river.

She was very happy at this and the monk carried her across. As they crossed, the old monk seemed quite happy and joked with the young maiden in his arms. Meanwhile, the younger monk strode through the water behind the couple, seeming distressed and sullen.

When they reached the other side, the old monk put the young girl down. They said good-bye, and the two monks moved away from the girl and the river. They walked for a while, and the young monk did not say a word. The old monk trod on in silence. But, clearly, the young monk was very upset. After about one hour had passed, the young monk finally broke the silence. "You took a vow of silence. You also took a vow not to be involved with any woman, especially a young and sexy girl like that. Yet you held her in your arms for more than fifteen minutes, and you seemed to enjoy it even more. I was shocked at your behavior."

The old monk remained stoic and said nothing. They walked on for another hour. The young monk was still troubled. He asked, "How could you break your vows that way? Won't you please give me an answer?"

Finally the old monk said, "I put her down when we reached the shore. You are still carrying her."

The story of the monks illustrates how identity with a false image conditions behavior. After leaving the monastery where their thinking was carefully conditioned by their vows, the two monks were nevertheless in different states of mind. The young monk, having created images of the older monk and himself as righteous carriers of spiritual rectitude who must follow the rules of abstinence, carried these images with him. The older monk, knowing full well that such a state of consciousness, while possible to maintain in the monastery where everyone had agreed to it, knew better of the "real" world.

While not even realizing it, the young monk had become "addicted" to his righteous image. He was practicing it without mindfulness. The young monk's addiction appeared as anger and resentment. Thus dukkha arises. This addiction commonly springs from feelings of unworthiness (a false picture of one identifying inappropriately with an image of another).

Perhaps false images lead to all forms of conditioned human behavior, both good and evil. But how do we free ourselves from false self-images? In the monk story the young monk carried the image with him for hours until the old monk relieved him of it by joking. Shamans also use many techniques, often involving the trickster image,[9] to acquaint us with our own often destructive, hidden, or unconscious self-images. When we see them, we are able to relieve or help ourselves understand our own suffering, and through effort and understanding, we are able to free ourselves from addiction to that behavior: in other words, free ourselves from our limited self-images and thus modify our behavior.

This doesn't mean that if you are addicted to a substance like tobacco and you free yourself from that addiction by recognizing that the root of the addiction was a false identity arising from a need to be loved or popular in school, that you will never smoke again. You may choose not to smoke again,[10] but once you see what the image is and what is at the root of this destructive behavior, you will no longer use the substance, material, or person you are addicted to in the same way. Most likely you will lose your desire for the addictive substance once the root is exposed to the light.

Similarly, many spiritual practitioners believe that once God wakes up to all of the suffering being created by human beings, He/She will give up

His/Her addiction to the material world. We might say that the vacuum will exercise self-control and stop creating by no longer bringing conscious matter into existence.

We Think, Therefore We Are All Hungry Ghosts

I believe we commonly suffer from a false image I call *personality disorder*. Buddha believed and taught that all suffering would vanish when one learned the truth of existence, namely, that we are far too restrictive in our thinking about ourselves. In an article entitled, "Are We All Hungry Ghosts?"[11] the author, Mark Epstein, explained that at the first cross-cultural meeting of Eastern masters and Western psychotherapists, the Dalai Lama was incredulous at the frequency of mentions of "low self-esteem." He asked several Western members if they had this and was taken aback when all of them said yes. For Tibetan Buddhists, such feelings represent what is called *the realm of hungry ghosts*, not human beings. According to Tibetan lama Sogyal Rinpoche, in Tibet a positive sense of self is cultivated early in life and generally supported by friends and family as a child matures. To possess a negative sense of self is considered foolish.

This negative image is also a common basis for illness and, once understood, provides a means for shamanic healing. Our suffering comes from a belief that we are what we *think* we are. When we realize that at another level, perhaps a mythic one, we are God-the-Universe-our-children-our-mothers-the-apes-in-the-trees-the-rocks-on-the-ground, when we see that we are all that there *is*, all suffering appears to melt away just as the boundaries separating our visions of ourselves vanish. However, a lot depends on how we see that image. To hold to it falsely just produces another kind of addictive behavior. Remember the story of the monk.

The Circle of Conditioned Genesis

The understanding of Conditioned Genesis is a most powerful exercise in Buddhist logic, a consideration of how interdependence differs from causal influence in the manner in which everything—matter, energy and suffering—arises all at once. These are synchronistically[12] connected, not causally or mechanically, as one might suspect. Here we will model the Buddhist form of Conditioned Genesis as a clock, with each hour representing one of the twelve factors leading to existence as we know it.

As I mentioned, the Buddha taught that suffering results from false beliefs held as truths. These beliefs can be understood in terms of Buddhist logic.

It goes something like this:[13]

> When A exists, B exists.
> With A arising, B arises.
> When A does not exist, B does not exist.
> With A ceasing, B ceases.

Looking at this carefully one is tempted to use ordinary symbolic logic as shown in Euler Diagrams to deal with these seeming syllogisms.[14] Thus one might write:

> If A exists, then B exists.
> If A arises, then B arises.
> If A does not exist, then B does not exist.
> If A ceases, then B ceases.

The common form of Euler diagrammatic symbolic logic could be pictured as follows:

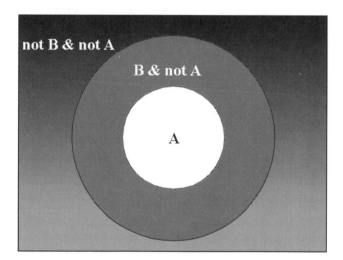

Figure 9.2. If A, Then B. This common form of reasoning applies to many situations, but not to Buddhist logic. The domain of A lies within the smaller clear circle. The domain of B lies within the larger shaded circle. Shaded circle B includes the clear circle of A and it also includes the area beyond A's domain, the shaded area of not-A and B. Both circles lie within the darker void of not-A and not-B. If something is clearly not-B, it is clearly not-A. But if something is clearly not-A it is not necessarily not-B. It could be B and not-A.

Here we see that condition A cannot occur unless it is within the boundaries of condition B. Condition B encompasses condition A. Thus if A arises at all, it must fall within the bounds of B. But it doesn't follow that if A does not exist, B shall not exist. For the area outside of circle A, known as not-A, overlaps with B within B's own boundary. This means it is possible for B to exist even if A does not, and it is possible for B to arise even if A does not. However, if B does not arise, then neither can A, as seen by the darker area surrounding both circles. Clearly if A is confined to the circle within B then anything outside of circle B must also be outside circle A. The logic that states, "If A, then B," equally assures us that "If not-B, then not-A." Thus, if B ceases then A must cease to exist.

If this seems obscure, consider an example. Let A be "All crows" and let B be "All birds." Then not-A would be the category "Not crows." This includes everything but crows. Thus not-A includes penguins, doves, tigers, and baseballs. Not-B would be "Not birds." This includes everything that isn't a bird. Thus penguins, being birds, even though they don't fly, would not be included in not-B. Tigers and baseballs would certainly be included as they are definitely not birds. So the statement, "If we have A, then they are B" reads "If we have all crows, then they are all birds" which is clearly true. But the statement "If we do not have crows, then we do not have birds" is clearly not true because the category "Not crows" also includes penguins, doves, and other species of birds.

We thus might be tempted to argue that the twelve interdependent conditions, called the *Patticasamuppada*, leading to suffering, follow as a chain of logic: If A then B; if B then C; if C then D, and so on. But Buddhist logic is not of this kind. It cannot be pictured as a nested set of circles. Instead all conditions must be placed on a single circle of reasoning. A, B, C, etc., are not separate. They cannot be put into the form of cause-and-effect thinking. They are one and the same. And so must be all of the twelve factors of the Patticasamuppada. Let's look at them as if they were hours on a clock face.

Again, we might be tempted to consider this as a chain of suffering. The metaphor seems unbreakable, with each link leading to the next link. However, when we break one link the other links remain intact. This implies that even though we were able to cause the cessation of a single condition, the others would remain linked together. But this is not the case. One link's break immediately causes them all to vanish. So, it appears to be better to regard the Patticasamuppada as a circle and the interdependent conditions as hours on a clock. Assuredly, as two o'clock follows one and as eight o'clock follows seven, the hands move in a circle, causing the next hour to arise so long as the one before it arose.

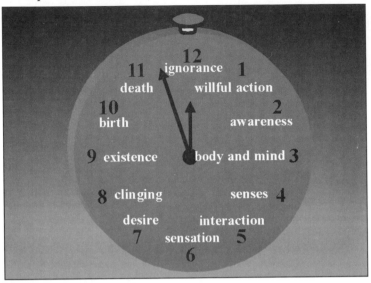

Figure 9.3. One, Two, Three O'clock Rock. We're going to rock around the clock, but time is running out. This is called Patticasamuppada—The Circle of Conditioned Genesis—and is responsible for the creation of the universe and all of its sentient creatures.

A key hour is that between 7:00, desire, and 8:00, clinging. To understand desire we need to grasp what its precedent is. Before there can be desire there must be 6:00, sensation. We can all agree on this. But what leads to sensation? All sensation arises as a result of 5:00, interaction. Something must interact with something else to produce that sensation. To have an interaction we must have 4:00, senses, abilities to perceive interaction. But to have senses we must possess 3:00, a body and a mind. I like to call this the *bodymind*, one word signifying the existence of physical and mental processes. Later I shall refer to this bodymind as simply *self*.

Here the Buddha's understanding of the process comes in. That which precedes the bodymind is 2:00, awareness itself. And what gives rise to awareness? The answer is 1:00, volitional action. And what creates willful action? Midnight, ignorance. Ignorance is a very special word. It doesn't mean not knowing. It means knowing at a deep level, but ignoring what it is that we know.

But why is there ignorance? The Buddha taught that ignorance naturally precedes the fact that at 11:00 we simply get old, ill, and die. In other words, death antecedes our ignorance. What creates death? Simply, there is death because at 10:00 we are born. Without birth, there is no death.

And why is there birth? The answer is a key to understanding addiction. It is 9:00, the existence of the universe, the creation of matter. But what comes before the universe's springing forth into material form? That answer is natural to all things. It is 8:00, grasping, reaching out, and clinging. This grasping arises from 7:00, desire. So you see, the whole thing is a circle. You can start the circle anywhere.

Now let's look at each of the hours of becoming in light of quantum physics: 1) Mental Forms Leading to Volitional Action and Karma; 2) Awareness; 3) Bodymind (Seat of Habits); 4) Senses; 5) Contact and Interaction; 6) Sensation; 7) Desire; 8) Grasping; 9) Existence: Matter Coming Into Being; 10) Birth; 11) Old Age, Illness, and Death; and 12) Ignorance: The Cause Of Illusion.

These twelve factors lead us to consider, in a new way, how compassion can free us from this cycle of life and death.

Midnight: Ignorance, the Cause of Illusion

Ignorance? What is it? In answering this, I will describe how the whole circle of suffering arises as illusion.

In the new physics we now understand that matter itself is unstable and is incapable of existing as our common sense tells us. Even ordinary and seemingly stable material objects soon lose their fixed solidity. The glass window you gaze out of is constantly flowing. The planet you stand on is continually changing, sometimes quite violently when shaken by tremors and plate tectonics. If we look at a rock in the water, it seems it will last forever. Of course it won't. Eventually the water will wear away the rock. And even the water changes form. It can evaporate and escape into the air and that air, consisting of molecules of many types, will not persist without changing form. The molecules can dissociate, turning into atoms. But what about atoms: surely they must be stable?

They are, to some extent,[15] but no one has ever clearly seen one. They are on the edge separating mental images from sensed objects. To grasp why they are stable requires us to take a new look at these most fundamental building blocks of the material universe we live in. Again, everything I say about atoms is based on a consistent theory of them, not on any observation of them.

The simplest atom is hydrogen. But in order for it to exist, the electron in that atom must be hidden; it must "dance" with the principle of uncertainty. From the vacuum's viewpoint, the soul, being the reflected consciousness of the vacuum, must remain uncertain of the electron's location. To accomplish this concealment, the *quantum wave function,* an invisible field that ripples the vacuum, must act on the electron by assuming a form (resembling an infinite honeycombed set of reflecting mirrors) that makes the electron appear as if it had an infinite number of positions at the same time.

None of these positions is real. Each of these positions is a possibility, but none of them would arise without the *intent* to look for them. Intent in quantum physics comes from the equations used to describe the physical situation. With each act of observation, these equations are altered as the quantum wave function is altered. Intent manifests itself as volitional action, a choice of observation. With intent, the vacuum of space ripples and polarizes due to the action of the quantum wave function, and the vacuum begins to communicate with the electron without knowing where it is really located. The electron dances with the cloud of possibilities by following its lead as if the cloud were *real.* The electron does not seem to lead in the dance. It just follows.[16] We might call this movement "ghost dancing" at the level of the electron. Think of the electron dancing in the dark with an invisible partner, its negative energy image contained within the Dirac Sea described earlier, and you have a picture. All electrons are haunted by vacuum ghosts.

Or think of this communication from the vacuum as a series of blows from an infinite number of tiny hammers, all hammering away, shaping the

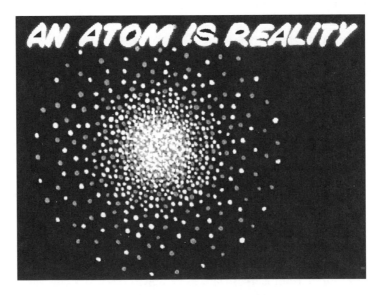

Figure 9.4. Atomic Reality. Each dot is a mirror reflection of the one electron hidden somewhere. There are an infinite number of dots but only one electron.

Figure 9.5. Dancing in the Dark with a Ghostly Electron Leading.

space around the atom into a cloud just as if a miniature sculptor were at work. This vacuum sculpting results in the atom taking a special form, reflecting the possibilities of the electron rather than the electron itself. In other words, the atom takes up a tiny volume of space and exists in time as a dancing cloud, one that becomes quite repetitive and stable. Without the vacuum and its action of intent, there would be no such thing as an atom, and all atoms would vanish.

If at any time the electron is actually discovered inside an atom, a so-called naked electron, the atom would become highly unstable. For the atom thus to be stable, ignorance must be present. The vacuum cannot know where the electron is really located. In that way the atom can exist as a material form. So, in a nutshell, because of the quantum world, there must be ignorance if there is to be anything at all. As it is at the vacuum polarization level so must it be everywhere else. As below, so above.

One O'clock: Ignorance Leads to Volitional Action and Karma

Let's look at the next hour in the circle. Volitional or willful action, or, as it is called in the East, karma, arises out of ignorance. Karma plays a major role in the process of reincarnation (as I will explain in chapter 11). Here we may think of karma as patterned response. Without ignorance, the mental forms, or "seeking," will not arise. Volitional action and karma are the possible mental forms that can arise in any situation. These forms are illusions, the seeking of stability, anticipation, fears, and desires. They correspond to our intent in the world, based, as it were, on data we have no access to.

In the atom, volitional action and karma are the imagined possible positions of the electron in the atom, laid out as a pattern called the quantum wave function. There is another way of taking action, another or dual mental form that leads to a complementary image, another possible performance for the electron: it can exist as a point particle, complementary to its appearance as the electron cloud. This, too, is a mental form giving rise to volitional action and karma. The cloud spontaneously appears when we choose to ignore the position. The position spontaneously arises when we choose to ignore the cloud when we actually attempt to measure the electron's location. This takes some doing and requires effort, which changes the boundaries of the system being examined.

Thus, we see a clue to how anything arises. By ignoring one mental form, a complementary form arises. There is no way to prevent the process from happening. The vacuum spontaneously bubbles and froths its dreams of matter into the nonvacuum world. And the vacuum, in choosing what to see, manifesting itself as the illusion of being an observer in human form, whether consciously or not, causes action and, therefore, karma.

Two O'clock: Volitional Action and Karma Lead to Awareness

Actions that bind or limit a system lead to awareness. These actions can be understood in terms of quantum physics. Without the mental forms, volitional actions and karma, there would be no such thing as awareness. Consciousness appears as a self-reflective field that constitutes the knowing of a mental form. Thus, when volitional action and karma are realized, awareness occurs, and, as we shall see, a possible and unconscious reality is brought into existence and awareness. None of this could occur without the action of reflection leading to awareness.

Here is where breaking the circle of addiction can begin. You simply have to notice the addiction! You need to become aware of yourself in the pattern that is addictive. So doing changes the attitude of reflection, half the battle. Once you are aware of a previously unconscious addictive behavior, the pattern no longer remains the same (hidden from awareness) and a complementary way of seeing is brought forward. The well-known Twelve-Step Program of Alcoholics Anonymous and other drug-addiction programs use this technique. When a person stands to make a statement, the first thing he says to the group is, "Hi, I'm Brian, and I'm an alcoholic."

When I spent weeks in the Peruvian jungle with shamans taking the vision-vine, ayahuasca, I saw my pattern of addiction. I, like so many Westerners with low self-esteem, was addicted to feeling sorry for myself. I refused to "clean up my act." I believed my illnesses were inevitable. But when I became conscious of my behavior, I was able to change that behavior. I no longer have low self-esteem.

Two-Fifteen O'clock: Peruvian Insight into Spirit

One night when I was with shamans[17] in Peru, I became very excited. I suddenly realized that trapped spirit, which results in bodymind consciousness, materializes and becomes the body. For spirit to be trapped it must reflect from its enclosing boundaries. These reflections build upon each other until they become conscious. I thought of it as the manifestation of a dream or the materialization of consciousness as a solid object forming from a dot pattern. I saw that the trapped spirit, as bodymind awareness, could become addicted to having a body, much as the body could become addicted to having a particular substance or other psychological dependency. As our bodies become addicted to material substances, our bodymind awareness becomes addicted to material form. This addiction appears as a desire to have something rather than nothing.

Thus, when we die, for a time we still carry a form of our material life, a dot-like pattern of our previous material addiction, and an outline of our egos. This shapes our karma in the next life. In this manner our habits from previous life experiences are shaped. We carry them into daily life, and unless we learn how to free ourselves from these habits, they become part of our surface armor, our egos.

Three O'clock: Awareness Leads to Bodymind (Seat of Habits)

It is very important to realize there is no world without a perception of that world. In other words, there is no universe without a perceiver of the universe.[18] If there were no consciousness, no means to realize mental forms, no ignorance, there would be no body and no mind at all! We would not be here, nor would we even know it. In that act of perception, that act of knowing, both you and the universe are born: that means the creation of an instrument that perceives or carries out the action of ignorance—volitional action and karma-consciousness. That instrument is your bodymind. And with this instrument, you perceive a direct experience of the universe as a duality. The whole universe, and everything going on in it, happens inside and outside of your body.

The bodymind arises from consciousness or, if you will, a desire for experiencing the volitional action and karma that arise out of ignorance. The bodymind is a battleground. Here we can see the direct forms that material will take as a result of the three previous hours in the circle: ignorance-volitional action and karma-consciousness-bodymind.

The bodymind is a wonderful instrument. It runs the show. It integrates both quantum and classical physics principles. The intent of the bodymind is to persist, to maintain a wide range of actions forming a seat of habits as an unconscious program. These habits become addictions because they are unconscious, hidden from the conscious will and thus uncontrollable unless made aware.

From ignorance comes habitual behavior, which is nothing more than the desire for stability. Without these habits there is no body survival. The body survives from the habits the bodymind has accumulated over millions of years of evolution. Only today are we beginning to grasp what the driving cause of that evolution is. It is not the past; it is the possibilities for the future—volitional action and karma.

Herein lies the core of addiction. The bodymind is that instrument that experiences the clock of suffering, the universe in miniature form, dividing it into a perceiver and a perceived. To perceive and to be a perceiver cause the problem, for the perceiver can never have the object of its perception. That object, the rest of the universe, is always "out there." Therefore the perceiver will always experience the lack of something. This is, of course, felt as desire. But to see how desire arises, we need to see what has come before the element of desire. The clock strikes four.

Four O'clock: Bodymind Leads to Senses

There is no realization of volitional action and karma, no consciousness[19] of them, without the senses. If the bodymind ceased to exist, the senses would vanish just as quickly. Thus, the bodymind is the seat of the senses. The senses are the windows that give view to the outer and inner worlds. The senses are designed to separate our spirits from the rest of the world. Through our six organs of sense[20] we both see the world *out there*, and, through our feeling compassion with other beings, sense ourselves *in here*. This compassion, felt *in here*, is the psychic awareness that overthrows the illusion of separation. If you feel compassion—the major piece of soul-evidence—your soul is present in your body.

Senses both tell us of the outside world and separate us from it. Senses are, in this sense, illusory, impossible without the body and the mind. Take away either and the senses become imaginal. They are there, but only in the guise of volitional action and karma, pre-bodymind aspects of the vacuum of space.

Senses tell us both what is out there in the universe, and because of how we experience those senses, that we are connected to the universe. Through our senses we know the *commonality of universal suffering*. We know pain and sorrow. We know pleasure and compassion. We know the soul.

Five O'clock: Senses Lead to Interaction

Let's look at what the senses tell us. It takes two to play the game of sensing. There is no sense without a sensor and a sensee. For there to be sensing there must be *contact*. In modern physics parlance, *contact* becomes *interaction*. When two objects interact, one object must "touch" the other in some manner constituting the interaction. Here we see that the senses lead to interaction. If senses were to cease from being, contact, forces, interactions, and energy exchanges would cease to exist, for there would be no way to know if they existed.[21] Without an eye you cannot have sight. Without an ear you cannot have sound. Thus the proverbial tree falls silently in the forest, if no ear is there to hear it. You can't have interaction without a separation of one thing from the other. So interaction implies distinction and separation and that is a clue to how desire will arise and another clue to which kind of addiction takes place.

Interaction is very important in physics. Through interaction, the various forces in the universe come into existence. We can see at the most fundamental level of awareness that empty space itself bubbles and froths into producing matter and antimatter from volitional action and karma—possibilities. This bubbling is the undercurrent of the universe. It gives rise to attraction. The matter and the antimatter, arising out of nothing, are attracted to recombine and become nothing again. When the universe first began, nothing became something in a very delicate upsetting of the balance of energy. There had to be as much positive as negative energy or else nothing could not become something.

Interaction leads to the desire to return to nothing again. Everything attracts. Even a repulsive force can be seen as desire, as a force out of balance attempting to return to a neutral state of nothing. The fact that matter exists is evidence that something is out of balance. Matter desires to annihilate with antimatter and return to the void from which it came. You can call this the death-wish if you will. You can also call this dukkha.

Six O'clock: Interaction Leads to Sensation

Without contact or interaction no sensation registers. Interaction leads to the direct experience of the sensations we experience and the desire to repeat them. Sensation leads to habits and the formation of unconscious actions we all take for granted. And how marvelous they are. We have a wide range of sensations: fear, pain, joy, sound, light, sex, cold, warmth, dizziness, enlightenment, and anger, just to name a few. These sensations are the very things we become addicted to. Indeed, without our sensations, we would know nothing about ourselves or the universe.

And what do sensations produce? The desire for a repeat performance. A young woman in the audience of one of my talks surprised me when I asked her what she wanted most out of life. She had a simple answer: "More. Whatever I have, I want more." Indeed, that is the marvelous thing about the whole circle of suffering. Every sensation creates the desire for repetition. The paradox of it all is that no experience can ever be repeated. Having a sensation the first time is quite different from the sensation the second time. That is why, for example, we are so addicted to falling in love. The first time was awesome. The next time is not so awesome. After ten years of marriage, ugh. It's just the same old thing. Why can't you change? I'm the same, I haven't changed. How come you don't make love to me like you used to?

As we find that every ah-ha becomes ho-hum, our sensations become habitual and more unconscious, so much so that the senses themselves are dulled into a stupor. The craving for a repeat of the first time experience never vanishes: that first cigarette, that first date, that first time you made love, that first time you fell in love. The mind craves a repetition. It never repeats. The unconscious formation of the habitual takes its toll and our sensations become duller.

Seven O'clock: Sensations Lead to Desire

A key insight here is that sensation gives rise to desire. When all sensations cease, desire is no more. All desire arises from the anticipation of a repeating sensation. Since a repetition is impossible, no desire is ever satisfied. Thus, the constant search remains to satisfy desire and causes addiction. There is no addiction if there is no desire to repeat a sensation. There is no sensation unless there is interaction. Interaction implies a separation, therefore, an object outside the person. The senses lead to interaction. And there is no sensing without the bodymind. The bodymind is the vehicle for the senses arising out of interaction, which takes place when consciousness acts, attempting to satisfy the mental formations of karma and will, that, in turn, pop up when ignorance manifests.

Desire appears as the tendency to repeat. Take a quantum system. The first time you determine its energy, you don't know what you'll get. Call this the "Forrest Gump" phenomenon of reaching into a box of chocolates. Once you measure it, even though the next energy measurement repeats the previous result, you, perceiving that information, are not the same. Ah-ha becomes ho-hum. While tasting the chocolate flavor, you have the illusion of its first taste of sweetness, but it never tastes as sweet.

Eight O'clock: Desire Leads to Grasping

When desire ceases, there is no need to clutch the world, no need to grasp and take hold. So desire is the anticipation of a repetition of the novelty of a sensation. Children delight in having a parent read the same story aloud over and over again. They seem to have the remarkable ability to repeat the same thrill they first experienced. Most of us have favorite movies that we watch again and again. Who hasn't seen *The Wizard of Oz* dozens of times? Yet each time we see it, we try to alter our perception so that it appears to us as if we were seeing it for the first time. Perhaps we don't quite remember each detail, but something triggers in us, and we find we anticipate the next scene, even though we may not really consciously know what's coming next. The unconscious mind remembers: "Gee, Toto, I guess we're not in Kansas anymore." All of this is clutching and grasping.

Nothing ever repeats; everything is different. Like children, we long to repeat any sensation that we enjoyed. We reach up and demand. We grasp. Desire causes us to do so. Every addict knows this. We all know what grasping is. It arises from desire and it requires action, which comes in the mind. Think of grasping as the mental projection, the karma of desire. Grasping is necessary for the self to be seen. Without grasping there is nothing called memory nor is there any way to bring forth a material world.

Nine O'clock: Grasping Leads to Existence— Matter's Coming into Being

Grasping is the mechanism through which matter comes into being. We can't have a universe without a grasping action. Grasping is turning a dream into a reality. It is the sudden collapse of a wave into a particle and the natural effect of something coming from nothing. When grasping ceases, the material world vanishes in a puff of smoke. Thus, to bring something into existence, there must be grasping. Through the desire to see a mental form take hold, we have the action of grasping.

Grasping causes material existence, which leads to birth. We see it in sexual desire. We grasp each other in sexual pleasure. The other person reaffirms our hold on reality. The other exists, therefore I am real. The result is, of course, birth. But the whole universe works this way. Every particle of matter grasps every particle of antimatter. The earth grasps matter to it. Gravity grasps. In quantum physics we see certain patterns of reality arise as a grasping of possibilities. Fluids form from the grasping action of molecules. Solids form from the grasping action of molecular and atomic bonding. Atoms form from the grasping action of protons and electrons. Nuclei of atoms form from the grasping action of protons and neutrons. And even they grasp quarks and other odd particles to form themselves.

Ten O'clock: Existence Leads to Birth

The result of existence, the attempt to fulfill desire, is birth. Birth is something arising from nothing. It is the creation of something new. What we fail to realize in grasping for a repeat of a past event is that every moment of existence is new. Nothing ever repeats. Nothing is the same. If we seek sameness and stability, we are caught by the circle of suffering. Thus, we will have desire and therein lies bringing all that exists into existence. Remember, the desire for stability comes out of ignorance that causes the whole thing in the first place. Recognize nothing is really stable, nothing ever repeats, and we can never fulfill desire. We grasp for straws, we seek illusion.

The Buddhists also see the action of birth as rebirth, a movement on the cycle of life and death.

Eleven O'clock: Birth Leads to Death

What follows birth is, of course, death. It would be a wonderful world if nothing ever died. I could live forever. But "forever" implies unchanging, and the universe is ever-changing. It goes on and on, never the same. It expands constantly. New things happen. *Death is the release of the illusion of stability.* This whole clock of Conditioned Genesis is also an illusion. Thus, because there can be no way to fulfill any hour on the circle, it is all fantasy. It is all a dream, a mistake.

As experience proceeds in the phase we call life, as matter exists in the phase called universe, illusion arises. Nothing is ever the same, yet we try to predict behavior, looking for cycles or patterns and for rhythms. In the *American*

Book of the Dead, E. J. Gold reminds us of the persistence of memory and the illusion of the material world. We live our lives so boringly because we fail to perceive that life cannot be experienced without recognition of all four phases of existence: birth, life, death, and transit. We are caught up with only half of the cycle, birth and life. But Gold maintains we must be aware, at every moment of our lives, of the reality of death and the transit stage.[22]

Noon: Transit Leads to Rebirth!

By recognizing that we're in transit (between death and life) while we journey from birth to death, thus learning something else is going on, we attempt to awaken from the big illusion that blinds us. Our job throughout life is to find out what it is! This attempt is not easy.

Since everything arises out of ignorance, something else is going on, something that is being ignored. You might say the universe is playing a trick on you. A few baubles and a few be-ables, and there you are, caught up with the show business of the universe. There you are, watching the show as if you were separate from all of the performers on stage. There you are, with your horse blinders on, watching birth, living life, and ignoring death as much as possible, and not even contemplating the life that exists between death and birth: transit.

Transit is difficult to realize because it is so ordinary in a weird sort of way. Take any everyday situation, and you can awaken to transit, if you dare. Take a trip to New Orleans. New Orleans is a transit city, especially if you spend any time in the French Quarter. Get on an airplane, lift off and you have entered the transit world with others. The other passengers, flying alongside you, are all aware, just as you are, of death. Welcome to the wake-up call of transit. Ever get up from sleep to take a shower and turned on the cold water when you meant to turn on the hot water? That moment when the familiar became *jamais vu* material was also transit.

Do you spend a lot of time alone? You are in transit. Spend a lot of time with strangers? You are in transit. Just get married? Transit. At a funeral? Transit.

We tend not to look at the weirdness of such experiences and, thereby, are unconscious of transit unless we make special efforts to become so. We are all too fascinated with the great wizard of our mundane life and ignore the little transitory guy in the corner pulling all the switches. Transit gives us a chance to realize something extremely important, but something that for many reasons we would rather not know. Or, put another way, the universe would like it very much if you did not know this. It hoists the flag that shouts Ignorance is bliss!

What do we all ignore and why do we do it? What is this something that we all, in the universe we experience, take for granted and have little awareness of or put little consciousness into? As Gold puts it:

> The Prime Aim when confronted with Reality is to realize that you are the shining void without losing yourself.[23]

At death you get a chance to awaken. Then all of this circle of suffering shows itself as an illusion, a dream of ignorance. There is no birth. There is no death. Isness just is. But what are we ignoring? Is there a hint of something that we do not desire? Is there a hint of something that seems to exist on and on without change? If it doesn't change, it is timeless. It must be outside all of the twelve hours. What is it?

Indescribable? Do you sense it as you read these words? Well, if not, then look at a hint of it. It is compassion. Notice I did not say love. As I see it, love is hopelessly confused with desire these days. Thus, no one loves today. Everyone loves tomorrow. They only desire to be loved. What we really have and do not need to seek is compassion: the realization of the suffering in the world. Compassion can give rise to sorrow and desire to do something. But whatever you do, if it arises from anything other than compassion, will result in nothing. Nothing can relieve the suffering because the suffering is itself an illusion. Nevertheless, we are caught in that illusion, it is powerful, and we experience suffering because we experience life itself.

What is the way out of this? The first step is compassion, the realization that all of this is suffering. Once you see this, you change. You may not drop your lowest forms of mindless activity, but it will be very hard to maintain them when you are aware of compassion. Compassion is only the first step to enlightenment. With enlightenment no addiction is possible, since there is the realization that nothing can ever repeat.

Of course, you can always go back and become ignorant again and ignore this teaching. You can drop it, saying, "What does he know? He is just a two-legged form like me." You can always ignore the truth. Even if what I say to you is not the truth as you would like to hear it, there is something in it you know to be true, that we all know to be true. For truth is that which needs nothing from this circle of suffering.

So what do I recommend? What do I offer you in conclusion? It's simple. Be your compassionate selves in whatever addictions arise. Remember that you are part of the whole universe and the suffering you feel is universal. It is necessary. It arises out of the desire to have something rather than nothing. It arises out of the need for stability. By remembering this, you will no longer be ignorant. And when you are no longer ignorant, the circle of suffering breaks and there is no more addiction.

This takes time, and it is not easy. There is lots of karma around these days. The addiction of spirit to material form is what we are attempting to deal with. When we lose that addiction, the whole universe vanishes. While we maintain it, there are many illusions confusing us. The trickster is always a breath away. What will we be left with? Or, should I say, what will I be left with?

To find out, discover the trickster who is hiding the universe from you by turning the page.

CHAPTER 10

The Soul Trickster:
Chaos, Lies, and Order

The fool who knows he is a fool
Is that much wiser.
The fool who thinks he is wise
Is a fool indeed.

—Dhammapada, *The Sayings of the Buddha*

The personality as we know it, the I that takes us through life
and stands in for us at work, in relationships, and in private
and public life, is based on lies.[1]

—Boris Mouravieff, *Gnosis, Book One*

ACCORDING TO GREEK LEGEND,[2] the god Hermes was the guide of souls. In ancient Greek stories Hermes is also the trickster: a thief, an inventor, humorist, and most importantly, the escort of souls to the underworld. According to Dr. Murray Stein, author and training analyst at the C. G. Jung Institute in Chicago, the archetype of Hermes was often represented by a pile of stones placed at a crossroads. In later representations it was built in the form of a quadratic pillar-with-phallus, symbolizing a connection of Hermes with Eros and placed at boundaries and entranceways.[3]

Why should Hermes, a trickster, be put in such locations? In many stories Hermes dodges across the boundaries separating worlds, particularly between the underworld and the earthly domain. Yet Hermes is also represented as the guardian of the underworld and is in general the guardian of all boundaries. Stein points out that this image indicates that Hermes is also the marker of delimited space and is placed at boundaries to provide direction.

In light of his role as trickster, this image of Hermes as director is especially astonishing. Why would any soul take direction from a con-artist standing at a crossroads? Hermes as the trickster-magician is represented in the Tarot by number two of the major arcana.[4] In the Tarot *two* stands for division or boundary-making and indicates creativity. Thus Hermes, as the magician, directs the flow of energy by making something apparently from nothing. Bringing both order out of chaos and chaos from order requires the trickster's prestidigitation, for the trickster is always upsetting the established order of things in surprising ways; thus, he is the creator of confusion as he crosses from the world of order into the world of chaos and the creator of order as he crosses back.

Ever since Job wondered about the Lord's deeds, we have wondered about our own precarious positions in the universe. The Tarot views the trickster as the clever fool or magician and as the principal of creation. Indigenous peoples throughout the world engage the trickster in one form or another during times of crisis and creativity. Why should reality during these times be so strange that it appears to be playing tricks on us? One might expect that the tricks of a fool would be the last things we need to see as we seek guidance. We most likely do not want to confront a deceiver at the boundary separating life and death.

Figure 10.1. The Trickster at Work. The trickster often encourages us to take a risk, a quantum leap, into something new.

Who Is the Trickster?

Apparently all of us have a great need to fool ourselves. The legend of Hermes asserts that he is the guardian between the worlds of life and death, and, in fact, the border-guard at the edge of all dualities. The legend teaches us that the answer to the question "Who is the trickster" is "You are!" Just as Hermes sits at the boundary between worlds, there also he places the seat of the soul, with one side of the chair in the flowing possibilities of the underworld and the other in the sensible world we take to be reality.

Richard Smoley, in an excellent article,[5] describes Robert Graves's story of King Philip II of Spain, who, during the height of Jewish persecution, decreed that every Spaniard with Jewish blood must wear a hat of a specific shape. That evening the King's fool appeared in court bearing three such hats. The King asked whom were the hats for. The fool declared, "One for me, Nuncle, one for thee, and one for the Grand Inquisitor." Philip got the point and rescinded his decree.

The trickster plays a role in everyday life, especially when people or nations are afraid to look at the truth and instead hold to a belief that no longer serves humanity. During such times, when the truth is not popular and often suppressed, the trickster appears as a clown joking with us, displaying his or her mischievousness. In the sixties when racial minority equal rights was a major issue and even talking about racial issues was offensive to some, the comedian Lenny Bruce acted as a trickster. Bruce was known for his scatological language, particularly racial epithets when his audience consisted largely of members of a specific racial minority. He repeated an offensive word continuously until it became a meaningless mantra and then explained to his outraged audience, a word is just a sound: Everything depends on how the audience reacts to the sound. He explained one should not let sounds hold power by attaching negative or fearful emotional significance to them.

The story "The Emperor's New Clothes" illustrates how the power to hold people in illusion through fear can be overthrown by innocence, such as the utterance of the child that the emperor was naked. Whenever there is self-deception, whenever pride or arrogance chokes off the life-flow of a human being, a family, a nation, or even a planet, the trickster seems to appear and, we might add, thank God she or he does.

Jung tells us the trickster appears whenever we tell lies and whenever we deliberately hide the truth from ourselves. He or she comes to us as the *shadow*, that strange and complex set of traits we all have but hate to admit to. Now since we all tend to deceive ourselves, probably more often than not, it is no wonder that the trickster manifests so often.

From what we have read in chapters 8 and 9, the trickster should be with us with nearly every step we take, since making a true statement is so difficult for most of us. Many people believe the trickster appears when we are most impressed with ourselves. But this isn't the only way the joker laughs at us. Any attempt to delude yourself provides an invitation to the trickster. Let me tell you about a period when the trickster appeared to me many times as a reminder to be patient with myself and that all crises I was experiencing would pass. The last time he appeared was in the guise of a one-legged bird.

The One-Legged Winged Trickster

The spring of 1977 in San Francisco was a crisis period in my existence. I was on the edge of some major life changes and was quite stressed out as a result. According to a modern theory of stress and its relationship to serious diseases, a person is most stressed and most susceptible to illness and accident when certain stress factors are present. These include (1) changing one's address, (2) changing one's marital status, and (3) changing one's job. If any one of these factors is present, watch out, illness or accident is around the corner. If two factors are present, illness or accident is nearly inevitable, and if all three are present, like the proverbial "three strikes you're out," you're practically on the edge of the grave, if you haven't already slipped into it.

I was batting with all three strikes against me and was especially anxious about the future when I found out that the Tibetan Karmapa, the spiritual teacher of the Dalai Lama, was coming to San Francisco to lead a ceremony for world peace. I'd received an invitation to attend and decided to see what it was all about.

The ceremony was held in a large auditorium on Nob Hill. Many thousands were in attendance, and I had a seat in the high mezzanine area, so my view, although unobstructed, was from some distance. As the audience came in and quieted down, the ceremony began. At first, the director of the institute who had made it possible for the Karmapa to visit came on the stage and told us this was indeed a very rare event, that the Karmapa rarely ever left his sanctuary high in the Himalayas to do this kind of thing. But these were rare times, and many felt the need to bring the East to the West in the form of spiritual teaching.

As the director was leaving the stage, a strange thing happened. The Tibetan spiritual teacher Trungpa Rinpoche, who then led the Buddhist Naropa Institute in Boulder, Colorado, suddenly limped into view from a position just offstage and behind the curtain. He said to the audience, very loudly and in a somewhat drunken slur, "Don't be fooled." He then limped off the stage and vanished from sight.

I asked my friend, "What was that all about?" She told me Trungpa was now drinking alcohol nearly all day, having suffered an automobile accident in Colorado that shattered his right leg. I thought I had noticed that along with his crutch he was carrying a bottle of booze with him when he came on-stage. Of course I knew the Rinpoche would have had some connection with his spiritual leader, the Karmapa, but nevertheless I was surprised to see him onstage, seemingly joking, and in the condition he was in.

But then the great long horns sounded and the ceremony began. It was as if I were watching a living *tankha*—a spiritual painting—come alive as the ceremony continued. The Karmapa donned his black hat and blessed the whole assembly. Many were surprisingly moved by the long ceremony. I was a bit bored. I wanted more action. As the ceremony reached a particularly high point of silence—perhaps better said, soundlessness (which for me was the height of boredom)—suddenly the drunken Trungpa once again, with crutch under his arm, limped onstage and said, "Keep your shirt on."

I turned to my friend who, being entranced with the ceremony, was not bored. I was about to ask her if she had seen Trungpa, when I thought better and decided not to interrupt her reverie. I couldn't believe he would have done it again, disrupted a sacred ceremony. I was feeling a bit anxious during the ceremony and laughed at the message. To me, keeping one's shirt on meant slowing down and being patient—a lesson I had not learned.

After the ceremony, as we were all leaving the auditorium, I asked my friend, "Did you see the Rinpoche?" She said, "Of course I did, don't you re-member? I told you so at the beginning of the ceremony." I said, " No, I don't mean then, I mean later when he came on the stage and said, 'Keep your shirt on'." My friend, rather astonished, looked at me and said she hadn't seen him come on a second time and she certainly didn't hear him say, "Keep your shirt on." I was a little mystified at this, but I attributed it to her being in a some-what mystical state herself during that part of the ceremony.

Later that week, while I was home in my flat in San Francisco, I received a phone call from a man I didn't know. He said he was calling me from San Jose, sixty or so miles south of the city. He had heard that I was interested in the Qabala and as he had some interest in the subject, he hoped to discuss the connection between the Qabala and quantum physics.

I was interested, but not knowing this person, I was a little apprehensive during the call. Nevertheless, I invited him to come to my apartment in San Francisco. He told me it might take him a little while because he didn't have a car and wasn't sure about the public transportation between the two cities.

I agreed to wait for him. After about three hours I was more concerned because he hadn't shown up. I took a peek out of my upstairs window and noticed a man coming up the street. He was visibly limping, had a crutch under one arm, and was having difficulty climbing the upward-sloped street. I noticed he had a very bad limp. He rang my bell and I let him in.

In about five minutes he was already out the door, acting very disgusted with our meeting. He said, "I went to a lot of trouble to get here and I find you know nothing." He was also a little intoxicated and that, coupled with the strangeness of his call and his bad leg, had me quite mystified. Had I received a visit from some clone of the Rinpoche? Whatever it was, I had to admit, in spite of the obvious discomfort of my visitor, it was strange and a little humorous.

I didn't think about it very much after that. Several months passed and my low self-esteem crisis had not eased nor had my impatience or anxiety. I was feeling quite despondent and still hadn't any idea about just what I would do to make my way in the world. Then, one day while I sat on a park bench looking down at the ground feeling even more depressed and anxious, a small wild bird with a broken leg hopped up to my shoe and pecked at it.

I had to laugh. The trickster had finally made it clear to me. As a one-legged life form it had to go slowly. First appearing as a drunken spiritual teacher reminding me to be patient during the long ceremony, then appearing as a drunken hobbling man, slowly climbing a hill, coming sixty miles to simply present to me a reminder that going slow did not mean not achieving one's goal, and finally as a one-legged bird patiently pecking at my shoe, the trickster had finally gotten through to me. It was too expensive for me to maintain my depression. By being patient the depression would lift. The chaos of my life would be brought to some order by my simply learning to wait. Whatever I was going to do, I needed patience. To this day, although I still have difficulty with patience, I usually slow down when I remember the patience needed to hobble along on one leg.

At the Boundary of Chaos and Order

The world of modern physics constantly deals with the issue of order coming from chaos and the reverse. To tell the truth, we don't quite know how this all happens. Our fundamental equations tell us everything is ordered, even though perhaps uncertain. But real life and real circumstances of that life are far from perfectly ordered. Error creeps into every situation imaginable. Einstein once wrote:

> As far as the laws of mathematics refer to reality, they are not certain; and so far as they are certain, they do not refer to reality.[6]

Reality appears at the edge between order and chaos. This edge is blurred, not sharply honed, and no one quite knows where to place such a fuzzy boundary. Mathematics and the laws of physics are marvelous inventions in bringing order out of chaos. They seem to be "out there," as real as Mars's elliptical orbit and Earth's spherical shape. They are as true as a pendulum's steady time-keeping period and as unwavering as the frequency of a laser's directed light.

But not quite. The laser's light frequency has some fluctuations in it, unpredictable and unwanted. The Earth's spherical shape is not quite real either. She seems to bulge more at the equator and has a distinctive pear shape perturbation to her roundness. Mars's orbit is continually being perturbed as well by the other planets and the asteroid belt, and even the pendulum's steady to-and-fro movement eventually slows to a nervous twitching. As I described in chapter 7, even that twitching is due to bombardment from the vacuum of space.

That interchange, that dance involving the movement from order into chaos and back again, seems to be the only way the soul ever really talks to us. In the ancient Chinese systems of thought, this dance, this energy, was called *ch'i*. Without ch'i, life is impossible and this entity we call the soul would never appear.

Dancing with Ch'i

The energy that is not really energy, ch'i, is actually the vital force of chaos interrupting our ordered and oftentimes dreary lives. Like a blackjack player at the gambling table calling for another card to come as close as possible to the goal of twenty-one, we need continual chaotic *hits* of ch'i to restore ourselves to as near a perfect balance as possible. We tend to get out of balance without hits of ch'i, for reasons that may not be apparent to us. Let me explain: The word *ch'i* literally means *gas* or *ether*. The ancient Chinese saw ch'i as breath in the body or the vital energy that animated it. Neo-Confucians, according to Fritjof Capra, developed a concept of ch'i that resembled what we in modern physics call vacuum energy.[7] Ch'i is thus a fluctuating field that looks like energy arising spontaneously out of nothing in continual interaction with the body and the mind.

It is not energy because if it were, the vacuum, even though it has a lot of potential energy, would soon be sucked dry, and the uncertainty principle would fail. Whatever the vacuum produces as energy, it just as quickly grabs back. Yet it bubbles and froths continually.

The aim of Chinese medicine is to balance ch'i in the body. It seems that through our everyday trials and troubles we find our ch'i stops flowing through parts of the body and gets stopped up, accumulating in some places and lacking in others. The goal of the Chinese physician is to find where the ch'i is blocked and, through the use of pressure often applied with needles to certain places in the body called acupuncture points, cause the ch'i to flow. Life itself causes the ch'i to circulate in the body, but every once in a while it needs help from a practitioner of acupuncture to get it moving around, especially when we are under great stress from work, relationships, lack of proper food or exercise, and poor thought patterns.

For example, suppose you have a common headache. To a Western physician, your pain may be caused by sinus congestion, but to a Chinese physician most likely your head pain results from too little ch'i in your head (meaning too much thinking or order) and too much ch'i in your stomach (meaning too much chaos or disorder). A needle properly applied results in a flow from the stomach to the head, balancing the ch'i and therefore restoring the body to health and no pain. Think of ch'i moving from one place to another as the flow of chaos.

Since ch'i is the continual interaction of all things with the vacuum, one is faced with two apparently different forces: *flows* of orderly information that provide direction and predictability and interruptions that breed chaos and uncertainty. Let us go farther in the analogy. Imagine that the orderly currents flow from both the past and the future into the present. The past attempts to provide us with the information, memories, rules, and rituals that we have come to depend on. The future attempts to provide us with anticipation, hopes, dreams, breaking of unmanageable rules and establishment of new rules, visions, and novelty. But then comes the inevitable chaotic interruptions of ch'i creating the *now* we feel as awareness. Like a peregrinating jester laughing at us, no matter which side we take in our stand for order, chaos seems to jostle our sensibilities. The nexus of this awareness is—you guessed it—the seat of the soul, the place where the two counter-streaming currents come together. *Now* appears as the internal and ever-present *consciousness*, ever hungry and ever searching for objects to pay attention to and in so doing jostling them about.

Ch'i is not the flowing currents from the past and the future. It is, instead, the continual interruption of awareness into that flow. Ancient Chinese philosophy suggests that the universe is made from ch'i, a form of energy that is in actual fact not energy at all, but, as we see, something more mysterious and quite removed from it. As the legend goes, the first ch'i is called Yuan Ch'i, the progenitor of chaos. Chaos was so upsetting that it divided into two parts, Yin and Yang. The soul has never recovered from this division, which

Figure 10.2. The Yin-Yang Symbol. According to Chinese legend, Chaos divided itself into Yin and Yang symbolizing every duality. As we see, within each side of the duality lies the seed of its opposite. Think of the seed as the trickster, ch'i.

is the fundamental act of creation or magic from which all trouble arises, including the male and female energy separation, the matter and energy space-time separation, and in general all duality. Instead, the soul is continually attempting to produce order, to get into the flow, or to correct itself by regaining balance, ultimately blurring Yin and Yang. But it is continually bounced back into chaos and again into the primal ch'i with every consciousness action. You might call this the karma of the soul.

One may compare figure 10.2 with Tipler's vision of the universe.[8] Even though chaos divided into Yin and Yang symbolizes every duality and every opposite, within each antithesis lies the seed of its thesis. Think of the seed as the trickster element. Although the edge separating the two pairs shown as a sinusoidal curve may be commonly held to be the edge of chaos, it is not. I suggest the seeds shown as two small circles, each holding the opposite pattern from their environments, contain the home of the chaotic trickster and the voice of the soul. Like Jiminy Cricket in the movie *Pinnochio*, the voice of your soul is a tiny but constant nudge.

It's a Joke, Son! (The Fred Allen Comedy Radio Show)

"It's a joke, son," is a line from *Fred Allen* radio shows when the "old-timer" began his monologues with Fred playing the straight man. The universe is meant to be funny. Is the universe not only an illusion but one with the intent of trickery at its very soul? Does the uncertainty principle mean God plays dice with the universe or is it that God somehow knows and causes the results of its seemingly random but not random movements?

Recent efforts in the field of mathematics have shown that even chaos has some order to it. Chaotic mathematical structures have been introduced to explain a variety of phenomena from the weather to the behavior of the stock market. Does chaos really exist or is this just another trick of our senses? The trickster principle implies that underneath everything is a deeper reality that shows our minds must play tricks with our senses so we may perceive, in a new way, an alternate order in the universe. The trickster element is the seed of the opposite in everything.

Remember that in modern physics, as physicist P. A. M. Dirac envisioned the void, particles of matter would spew forth from the vacuum, leaving behind holes that would behave as particles of antimatter. The holes can be thought of as the trickster element of matter. Whenever you have a creative thought, doesn't it always surprise you when the opposing thought also rises in your mind?

Trickster Balancing Act: Science on the Edge

Thank the trickster, that voice of your soul, for always attempting to bring things back into balance. Ancient Chinese martial arts and healing practices endeavor to teach humans how to regain their balance point to find *t'ai chi*. T'ai Chi means "great polarity" and refers to the concept illustrated in figure 10.2. Thus *ch'i* and *chi* are not the same words. Illness occurs when the ch'i (not chi) does not flow correctly. Ch'i and energy are to each other as perhaps mind and matter or as matter and quantum waves. These pairs are not truly opposite to each other, but are necessary requirements for there to be anything at all.

Western medicine uses the trickster principle to fool the body into getting into balance. Sufferers from lupus and similar diseases in which the immune system attacks the body find relief in medicines that are ultimately destructive to the body. Doctors offer these medicines to these patients along with non-harming substances such as cod-liver oil and rose-scented perfumes. The body is thus conditioned to believe that the perfume and the oil are essentially connected with the response of the body to the medicine. But the medicine is gradually withdrawn over time until after several years, the person receives such a small dose that nothing really preventive should be happening in the body at all. And yet the body maintains its balance against the disease. Researchers believe that the body is tricked into believing that the smell and the taste of the oil are necessary to fight the disease. Thus the body responds as if it were still getting the full dose of the medicine.

Candace Pert, former chief of brain biochemistry at the National Institute of Mental Health, believes the separation of mind and body is also an illusion.[9] She and her co-workers have discovered that neuropeptides previously thought to exist only in the brain are found throughout the body. Neuropeptides are known to be important in the flow of electrical activity, resulting in transmission of electrical signals from one neuron to another when emotional activity is going on. Thus, they are believed to be messenger molecules literally carrying information around the brain. Finding them throughout the body suggests that mind is present throughout the body and that emotions are the links between matter and mind or the mediators between them.

Trickster Oversoul of God's Dog (Spelled Backward?)

Since the soul within each individual often appears as the trickster, does anything like this occur to a society or perhaps to the world as a whole? In Native American traditions, such as the Navaho Nation's, the trickster appears during particular tribal rituals. The trickster/shaman dances and often acts the fool to remind the tribe to take an appropriate social action, usually one the tribe has been ignoring out of fear. Once the trickster has appeared, the people laugh and realize their collective folly.

Does the trickster also appear to our Western "civilized" culture when it is acting foolish? The coyote, although it weighs only about 30 pounds, is feared and distrusted by sheep ranchers in the United States and other countries. Yet, it is considered to be God's dog by the Native American peoples. They believe that to kill and skin the coyote releases its spirit and further upsets the balance of nature. To them it is as if we are killing a messenger from God. Perhaps we are.

The coyote is the trickster—the wolf we don't fear and the dog we can't trust—but has elements of both dog and wolf. The animal is bold and foolish, cautious and fearless, blending chaos and harmony. To some the coyote-trickster, existing in reality and in myth, plays it both ways, calling both heads and tails when the coin is flipped. The coyote teaches us it is a mature elder and a reckless child. It is a clown, a force of nature, and a messenger.

Where the coyote is a nuisance, the message is that something is out of balance. When we forget nature's delicate balance, the coyote appears to warn us, even giving up its own life to do so. Could the appearance of the coyote in Los Angeles in recent years be a warning that we have spread our flock too far into their territory and in so doing have put nature out of balance?

Most Native Americans believe this is so. Perhaps the trickster coyote is our messenger, our guardian at the boundary between order and chaos. Whatever it is, it certainly is giving us a message. We are all connected. Not just human beings, but all sentient life-forms. What we do affects not only ourselves, but others.

The lyrics of the Beatles' song reverberate in our ears, "Choking smokers, don't you know the joker laughs at you?" The trickster-messenger is us laughing at ourselves. Say hello to Hermes when you gaze into the mirror.

PART 4

THE FALLING SOUL:
OUT OF NOTHING WE COME

WE'VE PASSED THROUGH the eye of the Buddha's needle more easily than a rich man coming into heaven, according to Jesus' parable, and even though our souls don't exist as we commonly think about them, and even though anatta reigns in Buddhist logic, we shall fear no evil! No longer stuck in the materiality or nonmateriality of the soul, we now ask the right questions and find answers. We ask, like the annoying child in the crowd as the naked emperor passes by, "What happens to me when I die?" And now instead of shrugging our shoulders, we answer this question that reverberates throughout history. In reconstructing the soul, we find an answer. All the wisest seem to agree that we do continue after death even if they do not agree on what rattles around in the afterlife. Yes, we live again. Yes, yes, we continue on forever and ever until the end of time. And then what happens? Or is this all wishful thinking? Here we look at the question that haunts us, taking a quantum physicist's view—a strange and perhaps uncomfortable view, but nevertheless, the best, I believe, science can provide. You may not agree with what I see, but, then again, you might.

CHAPTER 11

Heaven, Hell, Immortality, Reincarnation, and Karma

The world of imagination is the world of eternity. It is the divine bosom into which we shall all go after death of the vegetative body.
—William Blake

Nay, speak not comfortably to me of death, oh great Ulysses. Rather would I live on the ground as the hireling of another, with a landless man who had no great livelihood, than bear sway among all the dead that be departed.
—Achilles to Ulysses, Homer

Sometimes before this in my dreams I have had a feeling that I lived again forgotten lives. Has none of you felt that?. . . Maybe life from its beginning has been spinning threads and webs of memories. Not a thing in the past, it may be, that has not left its memories about us. Someday we may learn to gather in that forgotten gossamer, we may learn to weave its strands together again, until the whole past is restored to us and life becomes one.
—H. G. Wells, *The Dream*

Some people call it Karma,
Some people call it guilt.
You can call it what you want to, honey
But I got it to the hilt.
—"New Age Blues," Lyrics by Ray De Sylvester, Scott Savage, Bil Thorne, and Chris West; music by Elfheim

PART 4 TAKES US OUT of the void and at the same time to a reconstruction of the concept of the soul. You could call it the physics of nothing-able-to-be-anything, a physics that you probably have grown to like, especially if you have read anything I've written.

In all of this soul-searching, the most difficult and incomprehensible concept to deal with is doubtless your own death—the idea that you will lose your I-ness. Think about it. Now you're here and the next minute you're sod. Or, maybe not.

Where Have We Been?

Let me first review where we have been in this book and briefly outline where we finish our exploration of a new vision of the immortal soul.

In part 1 we began our search by showing that if you ask the wrong questions about the soul, even if they seem rational, a wild goose chase ensues. Indeed, if you find answers you are left with a sense of futility. This, more than anything else, causes the unpopularity of the scientific method. Science itself is not to blame. However, its ground of being is false. Science is still based on a materialistic foundation wherein no experience *even exists* if it is not repeatable in a laboratory. Thus, a materialistically based scientific search for the soul is more or less doomed from the start. If the soul is not a physical, material substance, then it can't be contained by a materialistically shaped scientific enclosure.

Both Aristotle and Aquinas very early pointed to this difficulty by telling us that the soul's substance was so fine it lacked inertia and so could not be moved. Plato almost got rid of the substantial aspects of the soul altogether, somewhat mysteriously envisioning the soul as an eternal, ideal, and non-physical substance. Plato even showed us our attempts to define the soul were marred by our own souls becoming intoxicated when encapsulated in physical bodies. Thus came the mind/body or body/soul split, credited to Descartes, that we are familiar with and the dead-end at the close of our first search for the soul.

In part 2, we retraced our path, took a journey both backward and forward in time, literally examining nothing—the emptiness from which all things arise. We deconstructed the old materialistically based soul and, like the child viewing the emperor's new clothes or the fool of the Tarot deck, we found at the bottom of our deconstruction, aleph—the symbol of nothing-able-to-be-anything as the source of life, the universe, and the soul. But still the wrong questions persist in our thoughts. How can nothing become something? If it can, then why doesn't it dissolve into nothing once again?

An insight into the reason for our precarious existence came from the concept of the soul's war with time, symbolized in Qabala by aleph's struggle with yod—the soul in conflict with its reflection in matter. The timeless soul battles with the self, its own matter-based mirror image. The self's belief in the reality of this image, temporal though it be, often leads to violence, distrust, and all of the evils we know so well. The captured soul, forgetting its eternal root, follows a path of suffering and illusion as it attempts to satisfy a longing that can never be fulfilled.

Figure 11.1. Magritte's *Sea*: The Self Reflects the Soul as the Painting Reflects the Sea. The soul is like the sea, vast and indeterminate. The self is like the reflected sea in the painting, small, bounded, and always an imitation, an illusion.

Perhaps the Qabala leads us to ask how it happens that the potentially able-to-be-anything aleph, א, the fool of Tarot, the symbol of the soul-yet-to-be, creates a soul that incarnates and, having done so, manages to return that soul to the fool state once again.

According to ancient teachings the symbol of aleph must be drawn in a sacred manner. In fact, all Hebrew letters must be so drawn, with the writing instrument starting from the upper left corner of each letter to signify that we begin in heaven or the void. As the symbol signifies with its diagonal stroke coming down from the void, we seem to come from nothing. From birth, I have no memory of ever having been before.

As the symbol for aleph indicates, the bold lightninglike strike from the void into matter breaks the symmetry of the letter, so the lower left "leg," symbolizing duration in time, is separated from the upper right "hammer," symbolizing the beat of the intemporal. According to Biblical scholar Carlo Suarès:[1]

> The diagonal element of Aleph is thus the rupture of the inert crystal static state. It is the sign of . . . vibrations and wave undulations of intemporal and temporal, or discontinuous and continuous.

According to common belief, the soul never dies even though it passes from the body. Western religion is mute about the possibility of rebirth. But in Eastern religion, reincarnation is never questioned. Accordingly, I live forever and somehow reincarnate when I leave the vacuum of space and enter the world of space, time, and matter. Thus, having lived before, I burst upon the scene at the moment of my rebirth. Presumably, the shock of entering the temporal world removes any memory of where I've been or who I've been in past lives.

However, the return to the void seems to be a different kettle of fish. For, after all, on death I know that I have lived and have been mortal. At death I must enter the void with a full plate, so to speak, with all of my memories, regrets, and most regrettably of all, fears and hopes. Was I doomed to make those mistakes? Was it fate? Karma? Could I have led my life any differently from the way I had? If I continue after this life, how long will I last? Will I, upon rebirth, continue to err as I have in past lives?

Somehow, this universe was created and, according to some models, it will be annihilated. For what? Why such a ridiculous or majestic plan as this? If not planned, then how did chaos produce such apparent order? Perhaps the soul was created at the same time as the universe, and it will be destroyed when the universe ends. So nothing becomes nothing once again. But then what would keep it from becoming something all over again?

In attempting to define the soul as either material or not, we looked at the growing popular notion of a computer with consciousness. Instead of being matter-based, the soul is a program of instruction written in sand. Just how close sand-based units can come to being as conscious as carbon-based units, no one knows. I do know that something is missing in the sand, namely the essence of the spirit that blows it and gives it life. That something is *me*. Where am I in all of this?

In looking for the self, we returned to the vacuum and attempted to model the *soul* in terms of something physical, the sputtering energy of the vacuum and the spins of virtual negative energy electrons contained below the threshold of reality. Here the notion of the self being the soul's reflection in matter appears. We also saw, using a quantum physics model, how at the level of spinning electrons, self-knowledge and sacred or secret knowledge shape the way the world is perceived. Asking a wrong question not only leads to an unusable answer, but also leads to divisions, the creation of space and time walls and, in the soul's attempt to deal with the world, to soul-loss—the fall of the soul into the body.

There is an apparent paradox here. Most of us think of soul-loss as the soul leaving the body. What I am saying here is different, although it amounts to the same thing as is usually thought. The soul becomes so deeply submerged in the body that it loses its sense of the vacuum, its sense of the imaginal. It becomes crystallized or materialized. For the soul to be effective, it must maintain its imaginal connection. The self and the soul must be separated, not in space but in the quantum realm, as I explained in chapter 7.

Suppose the soul we seek is an illusion? Maybe I am an illusion. The Buddha said so in so many words. In part 3, we entered Buddhist territory where we had no soul or self to guide us. From Buddhism, I constructed three laws of emotion dealing with the addictive process of soul incarnation: inertia, action, and entropy. I showed how the usual form of logic we use is inadequate to handle Buddhist thought. However, the Buddhist way gave us an insight into the deeper meaning of good and evil and specifically showed us that in the fall of the soul into matter and its disconnection with the spirit, the self is left to flounder in illusions that I labeled addictions. The recovery from these addictions occurred when the self remembered who it was and reestablished contact with the nonmaterial soul.

If the soul fails to withdraw from the body during its lifetime, it becomes polluted and addicted. Thus the soul can become intoxicated while in the body. Ultimately, the pollution process leads to the separation of the soul from the self that we call death. Our souls are fooled by our senses according to Plato. This bit of wisdom is reflected throughout the shamanic world as the trickster, and, in the world of the occult vision of Qabala and medieval courts, as the magician or jester. Thus we were led to a new vision of ch'i as trickster-disruptions of flowing energy. Instead of being a flow of energy, it is the continual interruption of that flow by quantum leaps of awareness.

Where Are We Going?

In this chapter we will take a reassuring look at what heaven and hell are all about. They are not necessarily as they appear. Then, I'll explain what reincarnation is from a quantum physicist's viewpoint, why you will be born again, and all you'll need to know about karma before you pass from this mortal coil.

In the remainder of part 4, we face the difficult question of the soul's accountability and count-ability, that is, how many souls are there? I describe how you can talk to your own soul and provide some new definitions of spirit, soul, matter, and self.

Heaven and Hell

Christians throughout the Middle Ages put their faith in a literal interpretation of the Bible and as a result believed in the real estates of heaven and hell. Hell was widely believed to be in the center of the earth, quite a hot place since it consists mostly of molten iron. While medieval Qabalists and Muslim mystics taught that heaven and hell were to be taken symbolically not literally, Catholics, and after they emerged, Protestants, maintained their insistence on a literal interpretation,[2] a belief that persists today.

Brahmins believe in transmigration. A soul, atman, journeys from life to life in one reincarnation after another. The soul could migrate from person to person, from human to animal form, from animal to hell, from hell to heaven, just as a person moves from one house to another according to need. For Buddhists, transmigration is a result of conditions—causes and effects. Heaven and hell are imaginary creations of ignorant minds. One Buddhist disciple, Nâgârjuna, wrote in his book *Treatise on Relativity*:

> The hells are produced by imagination. Fools and simple people are cheated by error and illusion. . . . And the delightful heavenly palaces are also constructed by imagination.

In another famous story, Bodhidharma, the Brahmin I mentioned earlier[3] from southern India, who went on to teach Buddhism in 500 C.E. in

China, taught the meaning of hell to a disciple. An argument had ensued between a Chinese prince who denied the existence of hell and Bodhidharma who affirmed it. After the master patiently explained the existence of hell, the prince became so heated in his attempts to convince the great Indian teacher otherwise that he hurled insults at him. At this point, Bodhidharma patiently asserted, "Hell exists and you are in it."[4]

In the film *Jacob's Ladder* a Vietnam War veteran wakes up from a nightmare while riding the subway in New York City. It seems that something went wrong in "Nam." A platoon of soldiers was mysteriously wiped out, and the hero, Jacob Singer, remembers receiving a bayonet in the belly. Singer then suffers a series of bizarre episodes while living in the city. Although he is residing with a woman he desired when he was married to another, he continually flashes back to his married life and feels enormous sorrow when he pictures his youngest son, who was killed accidentally. Each flashback is terribly lucid, so much so that when he flashes, he feels as if he were really there. Even worse, he finds himself at a hellish party where a woman he's told is a psychic tells him he is already dead. As the story proceeds he suffers from being hounded by mysterious men in black who seem to want him to forget everything he ever knew about his platoon and what happened to them in the war. The only thing saving him from his nightmarish existence in the Big Apple is his chiropractor, Louie, who looks like an angel when Jacob glances up at him from the adjustment table while he straightens Jacob's back.

Jacob is afraid. After he tells Louie about his hellish adventures, Louie tells him that according to Meister Eckhart, hell isn't what it appears to be. All of those gruesome characters and all of those unpleasant memories that keep appearing in hell are only attempts to aid the soul in his or her journey by, if necessary, painfully tearing away the memories holding the soul to earth. Once the soul lets go, the hellish creatures vanish and the soul returns to the light. The devils are only there to help the soul on to the next life and into the light.

Jacob learns in the end that he is actually dead and his New York City transit experience was more than a subway ride, it was indeed *transit*—the movement from death to rebirth. In the final scene, Jacob is comforted by his already-departed son and taken by the hand up the stairs to heaven.

The story teaches us that hell may be more important for the addicted soul than heaven. Hell is painful because it frees the grasping soul from its clutch of the temporal world. In a sense, by falling into matter the soul loses its original boundaries, the alpha and omega points that mark the beginning and ending of time, and takes on the temporal boundaries of the individual bodymind it inhabits, the beginning and ending of a personal life.

In *Saved By The Light*, Dannion Brinkley recounts his visit to heaven in a near-death experience lasting twenty-eight minutes.[5] While on the telephone one day, Brinkley was struck by lightning and immediately felt himself out-of-body, rolling over in midair. He realized after an ambulance had arrived and the medics had taken a body on a stretcher that the body was his! He found himself entering a tunnel, or, as he put it, "the tunnel came to me." He then met a Being of Light (BOL) who engulfed him, and as it did this he was able to see, through time, all of his life from early childhood to the present moment. This also included, as in *Jacob's Ladder*, a period as a marine in Vietnam where he actually killed a North Vietnamese colonel.

When the BOL moved away from him, he felt all of his guilt leave him. Then the BOL took him upward where he heard humming and felt as if his body were vibrating at a fast rate. He saw energy fields, lakes, and views typically seen from an airplane. He was then swept into a city of cathedrals made of a crystalline glowing substance. He found himself in a room filled with loving light, and although he felt the presence of other beings there, he saw no one. Then several BOLs appeared, each with a box the size of a videotape coming from his chest, and in each box a vision of the future was presented to him. Later Brinkley was revived and remembered everything that had happened to him during his visit to heaven.

What are we to glean from these adventures in transit?[6] As I mentioned earlier, the time spent between death and rebirth is not to be taken lightly, or maybe I should say it is to be taken *lightlike*![7] Here I believe the habits that are to be formed for the next life are shaped as the soul looks at the past. Any unpleasantness, ghosts, goblins, and so forth are there to remind the soul of things it must let go of so the soul can move into the next incarnation with a lesson learned not to be repeated. Thus, hell arises during transit as the need of the soul to let go of all illusions.

There must be a number of flunkers in transit, considering all the suffering that persists. What causes this inertia of pain and suffering? The answer, it seems, has a great deal to do with the laws of physics, our understanding of time and our tendency to come into existence. I believe this tendency persists beyond the temporal world but once actuated as coming-into-existence immediately begins to lose its urgency. The tendency to come into existence is continually shaped and formed by those illusions that have not been released. In other words, hell isn't always successful. The polluted soul falls into matter sometimes carrying with it false information. Such information reinforces the way the soul will behave when it communicates with the bodymind. This is the effect of karma, and I have more to say about this later.

Immortality

Whatever one believes, there is popular interest in the soul, or something like it, that persists for immortality. Recently a number of books, films, and television shows have dealt with life after death, life in parallel worlds, and even living forever. The film and television series *The Highlander* portrayed a number of immortal souls who for some reason or another cannot die unless one is dispatched by being beheaded by another of them. Thus the series constantly deals with "our hero," an immortal facing one bad immortal after another. Presumably, the problem of immortal life is: How much good or evil can I accomplish?

If we look carefully for the common pillars of Western religious belief today, and this includes nearly every spiritual belief system, we find but two: the existence of some supreme being and the existence of an immortal soul. If we ask which pillar stands in every religious system practiced, God falls to the wayside while immortality stands, like its conceptual base, forever.

The concept of immortality includes a wide-reaching range of beliefs:[8]

1. Disembodied existence of a soul after death of the body;

2. Reincarnation of the soul, that is, continued existence in some other body;

3. Merging of the soul with the divine—a loss of separateness, a joining, in a non-temporal sense, with God, nature, or the universe;

4. Resurrection from the dead.

Playback of the Record at Transit Time

Given that we do go on in some way, consider what modern physics would say about this, particularly those first moments after dying. One can imagine a playback of the record of one's life at transit time, when the soul has the opportunity to consider what it had accomplished in life as it flees the body at light-speed or greater (superluminally). Modern relativity based on Einstein's theory shows that if an observer *were* moving superluminally, it could not be a physical object. Hence, the soul, if it moves this fast, cannot be made of matter.

According to ancient Druid teachings, at the moment of death, a lifetime's learning that has been built into the self is revealed. The body has recorded everything during its lifetime, registering everything on different levels. The mental, emotional, and spiritual planes have been recorded separately as if on separate sound or video tracks. At death several life reviews are conducted. Each review considers a different level of the person's life and is run backward through time until all levels—mental, emotional, intellectual, spiritual, and others—of the soul's existence have been revealed.

How could the soul see everything all over again and, in particular, running backward through time? According to modern relativity theory, if an observer were moving superluminally—faster than the speed of light—she or he would be able to see through time as if it were running backwards. She would witness events as if watching a movie running backward through time. However, according to relativity theory, the observer could not have any inertial mass; she or he could not be made of matter as we understand it. The individual would have to be an observer, capable of seeing, and not made of anything physical.

Perhaps witnessing is the wrong term to use. To witness something, he or she would have to see it, and that means she would have to experience it as light. But in going faster than light, he or she wouldn't "see" these events, but would sense them in some manner.

Figure 11.2. The Endless Cycle of Birth, Death, and Reincarnation. Seen as a cycle in the void, the cycle of existence takes the soul on many journeys throughout time and beyond. Meanwhile Shiva dances to the impossible life-death rhythm of aleph.

At death and during the early stages of transit the soul moves out of the body faster than light. This movement is like an endless circular ribbon of non-temporality (transcendence of time) winding its way through the atemporal void and then into temporal volumes of existence in different forms. These forms of existence—worlds of time—are the beats of the hammer of aleph. The beat goes on and on. This time around, as you incarnate, you get to play the king, the queen, or the joker!

Reincarnation

It seems we all return to the void upon exiting our mortal coils. But now we have a problem. How do we return as ourselves, as we were, and not as clean slates, *tabulae rasae*? If I am reborn, will I still have this feeling inside of me that says *I am I*, a living person with sensations, perceptions, mental images leading to my taking willful actions, and with the consciousness that I am doing this? Or, if I am not reborn as I am, what could ever be meant by reincarnation?

This is the big question of reincarnation. The answer is that you will retain the *I am I* feeling, but the reasoning is difficult to comprehend. The Buddhists tell us that thought itself can achieve nirvana; wisdom itself can realize nirvana. It is thought that thinks and wisdom that realizes. In other words, thought and wisdom are themselves self-referential, having a life and a sense of presence or "I"-ness of their own.

In Bernardo Bertolucci's film *Little Buddha*, Tibetan Buddhist monks search for the reincarnation of a dear and recently departed Lama. Their searches result in three possible candidates, all little children. Which of the three kids is the real one? To find out, they devise tests in hopes that a candidate will exhibit tendencies or habits remarkably similar to their lost one. Yet this proves exasperating. All three show some habits of the departed one, with each child showing a different facet or as the Buddhists would call it a different skandha[9] or aggregate, an accumulation of tendencies.

One of the children lives in Seattle with his parents, who, being Westerners, are relatively unfamiliar with such things as skandhas, as perhaps you are. They ask one of the monks to explain reincarnation. The monk takes out a candle and lights it. He says (as I recall the monologue):

See this flame. The candle will burn until it can no longer and then the flame will be extinguished. But if I pass the flame onto another longer candle, the flame continues to burn. The candles pass with time but the flame goes on. The candle flame is always the same flame, and yet it is always different. The candle is the body. The flame is what reincarnates.

Thus, a being who during life manages to free herself or himself from all fetters of desire, defilement, and other impurities can be compared upon death with the flame of a fire when the supply of combustible materials has been exhausted. It is the being that is extinguished, never being reborn.

But, what happens if that being never quite achieves such a noble goal? (This achievement is called nirvana.) In that case there is plenty of fuel around, and the being is, so to speak, reconstituted with many of the same tendencies and tasks ahead of her or him. And since the Buddhist monk who died hadn't quite made it to nirvana, the other monks were looking for where and when those particular idiosyncrasies would crop up again so the "soul" could be assisted on its way to nirvana.

But wait a minute, isn't this a contradiction? How can the monks help another soul when there isn't supposed to be one? Many students of Buddhism ask, if there is no **Self**, or soul, then what can realize nirvana? Put simply, if reincarnation exists, what reincarnates and why does it? The Buddhists answer that thought itself can achieve nirvana; wisdom itself can achieve nirvana. It is *thought that thinks* and *wisdom that realizes*. But before I can attempt to clarify these rather paradoxical statements we need to dig a little deeper.

Karma and Skandha

The key is in the Buddhist notion of skandha and in the Hindu notion of karma. I hope to clarify the conflict of the Buddhist nonsoul concept with Christian or Brahmin soul so that, at least from a quantum physics point of view, both Christians and Buddhists (I should say those who consider the soul real and those who don't) feel reconciled. In a sense, this resolution of an old argument is something like the wave-particle duality: To see the resolution one needs to step back from the fray and view it from a distance.

To Hindus, karma refers to action or doing. To Buddhists, it signifies the same thing but with a tiny codicil: only willful thoughts leading to actions are karma. Thus, with each conscious action, forces are brought to bear that lead to probable consequences. In these actions are both the identification of a self and the emergence of a not-self. Our world is made from such actions and their consequences. Mistrust and needless violence are with us today as a result of actions taken by our forebears that lead us to mistrust our fellow humans. But it is a game of probability, not one of determinism. It doesn't ever have to be the way it was. In fact, it never is the way it was.

Think of karma like the national debt: it is a powerful illusion. Our children and grandchildren could take it over after our contributions have been counted. Yet there is really no way to measure such a debt because it constantly changes, especially as our hearts and souls deal with reality and illusion.

Now, reincarnation implies that something that *was* is capable of changing form and becoming once again what it was in a different form.[10] Or something like that. It means something that was in carnal form, in a meat body but not meat itself, left that meat-tube, went somewhere for a while, and then returned to inhabit another meat-tube once again. The meat-tubes may differ, or they may have some DNA elements in common, such as a grandparent-child relation, or they may be as vastly differing as an ape from a mosquito, an apple from a Ph.D. candidate, a fig from a Newton. Yet something passes from one to another, and that something is real to those observers who knew both the original and the copy—the receiver and the sender of that something that passed.

The something that passes can of course be called the soul, and why it went from Newton to fig or from ape to Einstein is called karma. Well almost, but not quite. Karma is more difficult to explain than just that. We have all heard the statement, usually when something bad happens, "Oh, that was just her karma," as if it had everything to do with her unfortunate circumstance of being in Japan when the earthquake hit and the walls came tumbling down. Or when Joe Hardhat wins the lottery, we say Joe has good karma. Karma is something like that, but it is more than that, too.

To Hindus, karma refers to action or doing; in fact, the Sanskrit word karma literally means "action," implied by its root sound *kr*. It also implies duty or work toward an end. The Vedas use the word to denote ritualistic worship and humanitarian action. Shree Hirabhai Thakkar, a Hindu scholar, defines karma thus:

> Each and every physical action or deed that you perform with the cooperation of the mind, right from morning to evening, during the whole day and night, during the whole week, whole month, whole year and during the whole of your life right from birth to death.[11]

As we see, although it is usually not made specific in Hindu literature, karma seems to involve our mindful acts. In fact, according to Buddhists, karma means, rather specifically, willful action, not all action, and we shall get to this later.

Reincarnation and karma can be viewed as virtual vacuum energy processes utilizing an essential action called skandha—an aggregate of tendencies to come into being. Just as electrons maintain their character as they bubble from nothingness into the physical universe, so do the essential actions of the soul. As such, they result in reincarnation without the "I." The "I" is only a convenient label. The persistence in the tendency of these often aggravating aggregates to come into existence is real. These aggregates are like habits of the vacuum. They tend to recur in a person, a self, or an incarnated soul.

Five Scandalous Skandhas

I mentioned earlier that at the heart of the Buddhist notion of dukkha, which popularly means "suffering," is the subtle notion of the appearance of the "I." The Buddha referred to the "I" as nothing more than five existential tendency-forming skandhas called *pancakkhandha*, literally meaning five aggregates: matter, sensations, perceptions, mental formations, and consciousness.

Here we want to examine these aggregates more carefully to see how the Buddhists could regard reincarnation and the nonexistence of the soul simultaneously.

Matter

We begin with matter, the first aggregate. For ancient philosophy, including Greek and Buddhist thought, matter was composed of four elements: solid or *earth*, fluid or *water*, heat or *fire*, and motion or *air*. Six derivatives, all associated with how we distinguish these material elements, joined the elements. These include the five sense organs we possess to perceive the different phases of matter (eye, ear, nose, tongue, and skin) and the corresponding objects of these organs: visible forms, sounds, odors, tastes, and tangible or touchable things. To this physical list of five the Buddha added a sixth derivative: the mind and its targets, called mind-objects. Although the Buddha assumed the mind was nothing more than a sense, no more special than taste or smell, the mind nevertheless dealt with something quite different from the other five. What was the essential difference? I'll tell you in a moment.

Today our basic concept of matter has changed very little. Instead of seeing heat or fire as elemental, we view radiation consisting of light, X rays, infrared and ultraviolet, and so forth, all composing the electromagnetic spectrum, as part of matter. We recognize that our bodies are sensitive to these forms of radiation, and, in fact, we use them to look into the body and diagnose diseases.[12] Water and air are also different forms of matter, in liquid and gas form. We recognize matter in four phases: solid, liquid, gas, and plasma (ionized gas). We could add other phases such as molecular and atomic, subnuclear, and quarkian—this last one referring to the nuclear constituent particles making up a single nucleon.

Sensations

Before I go into this *matter* in more modern detail, let me complete the Buddhist list. The second aggregate is sensations. These correspond to the six derivatives in the first aggregate of matter. These six sensations include all sensations, those that both please and disgust us; they are the means by which we determine there is an "out there," out there. Remember that to the "normal" five senses (seeing, hearing, smelling, tasting, and feeling) Buddhists add the sixth sense: thinking/visualizing and the like—the sensing done by the mind.

Here the difference between the five senses and the mind-sense shows up. The five sensations dealing with the "normal" senses form the *spatial senses*, those that tell us space outside our selves exists. The "normal" sensations arise when these spatial senses are ignited. From these five "out there" senses we determine that space itself exists, separating each of us. The sixth sensation, mind, deals with time. Thus, for example, a thought is a sensation arising in the mind-sense, telling us about time, while the odor of garlic is a sensation that arises in the nose-sense, telling us about space.

The Uniqueness of the Sixth Sense: Mind Objects and "I"

I see mind as the birth of the temporal sense, the sense of time so intimately connected to thought and my model of the soul. Much as in the special theory of relativity, in which space occupies three dimensions and time the fourth, an imaginary-spatial dimension, mind takes up the imaginal dimension of time while the other five senses take up the sense-based or physical dimensions of space.[13]

Here the major distinction between what we call the "I" and what the Buddhists call the "not-I" arises. For most non-Buddhists, including Hindus, Christians, Jews, and Muslims, there is a soul, capable of eternal existence. We normally begin to sense the presence of something called soul or, if not soul, at least self, and, if not self, at least ego, with the temporal sense of mind arising in response to and simultaneously with the five spatial senses of the body. The mind-object here becomes this mysterious "I" we remember. We see it when we look back at our childhood photos, when our parents held us in their arms; we remember this self when we look at photos of special events in our past life, like marriages, births, and deaths of loved ones; we remember our selves when we watch old movies or hear the voices of old friends with whom we have not spoken in years.

Sweet though our memories may be, to the relentless eye of the analytical Buddhist mind, all of this is a door swinging in the wind, empty of substance, an illusion. Mind is a sense, and mind-objects are the illusionary foci of this sense. The mind senses something "out there" or, if you wish, "in here," but something real, as real as anything is considered to be real in Buddhist theology. As an event is marked by spatial dimensions and time, so is a memory marked in time by the other spatial senses.

Ideas and thoughts are also part of the real, tangible world. Yet while objects detected by the five spatial senses can be experienced with the mind sense, mind-objects cannot be sensed with any of the other five senses. We cannot smell or taste an idea, yet an idea can bring any of the other five senses into play: for example, the sensation of smell or taste to our noses and tongues. You don't believe it? Think about putting a tart lemon in your mouth and you get the idea. More than getting it, you probably taste it as your mouth puckers and waters. In a lucid dream, I have had the experience of reaching out and touching an object and feeling it as tangible in my hands. As I awakened, the object disappeared.

In fact, the above ability cautions us that the mind cannot sense anything different from an object sensed by one or more of the other physical senses. Mind-objects are made from the experiences of the spatial-physical world. That lemon you imagine in your mind, the color of yellow you see, the feeling of its weight as you hold it in your hand, the sensation of its juices running over your fingers as you cut it with a knife, the tart, bitter taste of it as you place it in your mouth, all of those sensations, as strong as they may be, are only mind-objects conditioned by your physical senses.

If born blind, you cannot perceive yellow in your mind. If born deaf, you cannot cannot know what Beethoven knew before he went deaf (and what he was still able to hear as mind-sounds after his loss). With nothing coming in to the other five senses, nothing appears in the mind.

Perceptions

The third aggregate is perceptions. Again conditioned by our senses, our perceptions fall into six sensorial camps. But perceptions are related to memory—the ability to cognate and put into perspective what our senses provide. Yet we do not really know how we are able to perceive anything. Sensing things is not a problem. We know how to construct a simulacrum, a sensory machine, that will be able to measure physical attributes as we imagine our five spatial senses do. We know how to construct a robot that will smell the air, feel heat, sense weight, feel pressure change, measure or determine the molecular composition of anything at all.

I mean this when I say we know how the senses work. But what about perceptions? To perceive something, you must recognize an object. How can I construct a machine that recognizes things? That, it turns out, is so difficult to accomplish that no one has done it yet. What does it mean to recognize something?

But don't computers recognize objects? Modern computers installed in artificially intelligent robots seemingly exhibit abilities to recognize the existence of objects and move around them, pick them up, and place them somewhere else. They can even alter an object, making it better than it was. Isn't that object-recognition?

Consider this machine I am using to write these words. It is a modern computer using a word-processing system. This system can be said to recognize the words I write. I can tell that it senses them, because it is capable of telling me how many words have been written in this document. The program acts like a mind, weighs the document, attributes a weight-quantum to a single word, and tells me how many words are present. My computer's word-processing program even recognizes a misspelled word. Does the program perceive the words I write? That is more difficult to say. For *exmaple*, perhaps it does since it can correct the spelling of any word I write within *certina* bounds. No, don't correct the spelling mistakes shown in the italicized words! Leave them there: *exmaple* instead of *example*, *certina* instead of *certain*. The program can check and note that these are misspelled words. It does this by asking me if instead of *certina*, I meant *certain* or *cretin* or *Cretan*, words it contains in memory and can substitute for *certina* at my will.

Does the program perceive these words? I don't think so, yet memory is necessary for the program to work. The program has a dictionary and is capable of quickly checking words that have two letters commuted or a letter missing. Then it checks all words that sound like the word written. It does this, not by perception, but by lists held in memory to which it can compare any word I type.

Perhaps all a perceiver needs to do is compare sense-objects with lists of sense-objects in memory and decide which object from the list is the closest match. But what decides closeness? And what enables one to recognize a sense-object from a mind-object? What tells you that you are dreaming of sucking on a lemon and not actually puckering your palate? Could a computer suck on a lemon in its mind and pucker its mental mouth?

I don't think so. But, if it is true that thought thinks, that is, that the be-ingness of thought implies a doingness of action, then the Buddha was right. The computer does make thoughts and since thoughts are thinkers, it should be able to pucker its mental mouth and experience sourness as you and I do.[14]

Mental Formations

This would imply a computer could have the fourth aggregate, mental for-mations, ruminating in its memory. But could it really? According to Bud-dhist thought, mental formations are the same as willful actions. To grasp this, think of mental formations as arising from wishful thinking, even if they are seemingly arising unconsciously. Daydreaming about that special vacation, new car, money, new sex partner? Then you are engaging in mental forma-tions. Here we get into trouble and find karma rearing its head. Whenever a willful action takes place, according to Buddhist thought, the law of karma comes into play. I shall return to this in a while, so let me go briefly into the fourth skandha without getting into karma per se.

Just as matter, sensations, and perceptions are of six varieties, so are there six kinds of willful actions. Thus there are willful actions associated with each of the five body-senses *and* with the mind-sense. For the Buddha mental formation means all willful actions, all volitional activities. Volition is always mental activity produced by mental formations. While not all ac-tions lead to karma, all willful and therefore conscious actions do. So far this aggregate does not appear to be present in computer artificial intelligence. Hence a sour lemon would have no taste for a computer because it could not intend to taste it.

Consciousness

Let me now go directly to the fifth aggregate, consciousness. Consciousness, for the Buddha, is a material phenomenon, a reaction or a response based en-tirely on the six senses, and nothing more. Consciousness cannot and does not exist unless it is responding to an object.

Although books have been written about consciousness without an ob-ject of consciousness, so-called pure consciousness, I have never been clear about what that means until I considered how Buddhist thought regards it. Consciousness itself does not recognize an object. Consciousness is just aware-ness of presence without discernment of detail. Thus, when you see a blue ob-ject, when the eye comes into contact with light-rays reflecting blue color, visual consciousness arises indicating presence of color but not blueness.

When the third aggregate, perception, comes into play then blue is seen as blue and recognized as such.

When you hear the melodious voice of your sweetheart or your mother calling you to dinner, consciousness does not recognize your mother's or your sweetheart's voice, only that sound is present. Perception comes into play to bring the vital quality of whose voice beckons.

When you taste a pork chop covered in brown gravy or a slice of hot apple pie, your consciousness does not know what you taste, only that something is being tasted. Perception again rings in the savory pork or the cinnamon spice flavor of the apple pie. Ditto with smell and touch and ditto with the sense of mind, and here is where the trouble I hinted at above begins. For the mind-sense also only knows a mind-object is present, and it does not discern what the object is. The object is indeterminate until perception occurs and then the mind-object is perceived. Thus consciousness is an *imaginal realm* phenomenon that arises in time, as mind-sense, whenever an appropriate action occurs in either the physical or imaginal domain.

I shall show how these five skandhas create the semblance of a unique being, but first I need to return to the fourth aggregate, mental formations, to deal with karma and will.

Quantum Karma Skandha

O, Bhikkhus [Monks], it is volition that I call karma. Having willed, one acts through body, speech, and mind.

The Buddha said those words a long time ago, and in them we find an essential truth that can be explained with the help of quantum physics. As I pointed out earlier, in Sanskrit karma refers to action, any action, while for Buddhists, karma refers to actions taken with will or volition always connected with mental formations, or, if you will, the construction of mental forms. In Hindu culture all actions create karma, while for Buddhists only willful actions do. Will directs action in the same way a film director directs the actors in a screenplay.

Just as there are six senses, six perceptions, and six forms of consciousness, there are six types of willful action. These are found in all forms of willful action including, but not limited to, attention, will, determination, confidence, concentration, wisdom, energy, desire, repugnance or hate, ignorance, conceit, and self-image. Basically, if you create an image which results in your taking an action, you have set the wheel of karma along its inevitable track and put the axle out of joint at the same time (dukkha).

Will is seen as will to live, will to survive, will to continue, will to increase or decrease, will to have more or have less. Strongly connected with this will is desire. Here we find what is termed the *second noble truth*, the arising of dukkha and its cessation. Such desire conditions or sets up willful action. Without desire there simply is no will or volition.

I explained in chapter 9 that spirit exists in the pure vacuum state and desires to become something rather than nothing. Thus the vacuum has a great urgency to spew forth energy and, ultimately, matter, a process I called vacuum compulsion. Call this the spiritual necessity. On top of all this, the soul, being fundamentally spirit, feels this compulsion similarly to a person addicted to a substance and in need of a fix. So while the spirit desires to bring forth matter, the soul, arising in the vacuum, has an exigency: a need to bury itself in matter.

While similar, spirit and soul are different. For now take it that spirit is unmanifested vacuum vibration, while soul is the same thing but reflected back on itself so a pattern within it can be seen.[15] The soul is able to manifest intent through this reflection, while the spirit is not. The spirit desires but knows not what it does nor even what it desires when it spews out matter and energy. The soul desires and knows what it is about and what it craves. The spirit is potentially conscious *and* potentially unconscious, while the soul is actually conscious *and* actually unconscious. (I'll have more to say about the relationship between spirit, soul, matter, and their connection to self in chapter 14.)

Thus, one of the major differences between artificially intelligent mechanical devices that seem to recognize objects and a human being is the means by which will or volition lead to or connect with mental formations. According to my model, based on quantum physics principles, will manifests as intent—the desire to manipulate any situation for some gain. Intent always upsets the apple carts of existence, brings karmic action into play, and generally modifies the behavior of objects under its scrutiny. It would seem that desire and intent are soul-properties and do not exist in matter alone. They can be embedded in the body, however, if a self is present. In order to grasp what occurs when we desire something we need to look at the following . . .

Psychoid Physics and the Vigilance of Intent

In my picture, quantum mechanics leads to a set of rules governing the psyche or the imaginal realm and the means by which that psyche controls the physical space around it.

I wish to consider two aspects of will here. The first addresses how will manifests in the physical world. The second considers how the soul communicates its desires to the self. Surprisingly, the answer is the same in both accounts: through vigilant observation. But the difference is between what constitutes the partners in the communication, that is, which is the observer and which is the observed?

How is it that what I wish to accomplish sometimes occurs without seeming effort, while at other times, even with great expenditure of energy, I fail in my endeavors? According to quantum physics principles, observation and awareness have a far greater effect on the physical world than was previously suspected. But what ties observation and awareness to the physical world? None other than desire. What you may not be aware of is how desire manifests as a physical force.

Desire first appears in the mind as a mental formation, the object of desire. When this form manifests, desire holds or preserves it in mind. The power to hold the image, determined by the time-span the image is held, is called *intent*. Next, the object that appeals to our senses is sought in the physical world if it is not present, and mind-object and the physical object are brought together and superimposed in the mind-space in much the same way you fuse two two-dimensional images together, making one three-dimensional image when you look at an object using your two eyes.

Desire, through our powers of observation, actually modifies and alters the course of the physical world, causing things to occur that would not normally occur if they were not desired.

The Quantum Watched Pot

How does desire affect the physical world? Intent operates in the physical world by altering the observed state of that world. The fact that intent affects the physical world reflects a recent discovery by quantum physicists Yakir Aharonov and M. Vardi that has received experimental verification by Wayne Itano and his colleagues. Aharonov and Vardi have shown that the old proverb "a watched pot never boils" may have a range of validity previously unsuspected. They have discovered a paradoxical situation that arises when a quantum system is watched carefully. As they put it:

> If one checks by continuous observations, if a given quantum system evolves from some initial state to some other final state along a specific trajectory . . . the result is always positive, whether or not the system would have done so on its own accord.[16]

If a quantum system[17] is monitored continuously, we could say vigilantly, it will do practically anything. For example, suppose you are watching a quantum system in an attempt to determine just when it undergoes a transition from one state to another. To make this concrete, think of an imaginary subatomic "quantum pot of water" being heated on a similarly sized stove. The transition occurs when the water goes from the calm state to the boiling state. We all know pots of water boil, given a few minutes or so. You would certainly think the watched quantum pot would also boil. It turns out, because of the vigilant observations, the transition never occurs; the watched quantum pot never boils.

Another example is the decay of an unstable system. On its own the system would decay in a few microseconds. But if it is watched continuously, it will never decay. All vigilantly watched "quantum pots" never boil, even if they are heated forever.[18] The old proverb of the watched pot thus turns out to be true in the quantum physics domain. What follows is the physics of karma (action of intent) consisting of three new quantum axioms of intent based on this proverb:

1. A watched quantum pot never boils if you *intently observe* it to not boil.

2. A watched quantum pot boils if you *intently observe* it to boil.

3. A watched quantum pot boils even on a cake of ice if you *intently observe* it to boil.

This implies there is a deep connection between the observer and the observed. So deep, in fact, that we really cannot separate them. All we can do is alter the way we experience reality. This is where intent comes in.

If the system were unobserved, it would certainly undergo the physical transition. The pot would boil. The observer effect causes the anomaly to occur. Let me explain. When the system is first observed, it is seen to be in its initial state. When it is observed just a smidgen of time later, well before the time in which it should change, the system is observed with more than 99.99 percent chance to be in its initial state. In other words, the system is found to be exactly where it was initially. Now repeat this measurement again and again, each time just a tiny bit of time later, and with a very high probability, the same observation occurs: The system is found in its initial state.

But time marches on, and eventually we pass all reasonable time limits for the transition to occur, yet it still doesn't happen. The system "freezes" in its initial state. The only requirement to freeze the motion is that the observer must have the intent to see the object in its initial state when he looks. This intent is determined by the frequency of his observations. He must look and find the object in the same state repeatedly in very short time intervals. Eventually a longer period of time passes.

We might question the physicists as to their mental intent in doing this experiment. We don't have to. By observing the system as they did, their intent was already established, already "out there," regardless of what they were actually thinking at that time. In other words, their intent was already a physical manifestation determined by the frequency of their observations. The old adage with a twist: you will see it when you believe it.

Suppose a physicist doesn't watch vigilantly or suppose that she or he does but with the intent of seeing it evolve naturally. Then what? Take the quantum pot. If the physicist looks intermittently, expecting it to boil eventually, the pot will follow its natural unobserved course and will boil as proved. These observations, because they are infrequent, have little effect on the natural result. Or if the physicist wishes, she or he may observe the object vigilantly along its natural evolution, and will observe the same result. In other words, a watched pot boils if you intend it to.

Finally, there is another bizarre element to this. Suppose the system could be observed to evolve along a bizarre path, a highly improbable mission, so to speak. If the intent to observe *that* occurrence is vigilant enough, the object will actually follow the bizarre path of evolution. You can make things happen simply by intending them to happen if you observe your intention with great vigilance, intense observation occurring over very short time intervals, more or less continuously but along a new, unexpected track.

I need to caution the reader here. The bizarre path of evolution seems strange because it violates the second law of thermodynamics: It seems to move energy from a cold to a hot body without performing any work. Although such a path of evolution does not violate the law of conservation of energy and, therefore, is possible, it certainly is not very probable. This is no problem in quantum physics; even the most improbable occurs once in a while. The observer here is vigilantly watching for that rare occurrence and ignoring by not looking for the normal path of evolution.[19] Hence, a watched pot boils on a cake of ice, if you intend it to.

I need to point out that intent and intentions are not the same. Intent refers to a vigorous action of vigilant observations along a specific path of evolution. It matters little what you hope for or even what you passively expect will happen. You need to actively pursue your vision to manifest intent in the physical world, not passively dream about it and hope it will come true. The direction of evolution is determined as you go and depends on where you focus your observation. Thus, intent requires a quantum physics basis. Intention, on the other hand, is a classical mechanical concept. One sets in motion a certain expectation and then hopes for the best. The old adage "the best laid plans of mice and men often go astray" tells the whole story.

Our brains may be composed of like quantum systems, and consequently our paths through history may be governed by the pot-watched-with-intent theory. Thus, this may be how will and intent actually govern the movement of living sentient systems.

If There Is No Soul, Then What Reincarnates?

We have seen how Buddhist aggregates operate in the creation of both physical objects and the five senses to relate to those objects in the three dimensions of space. Similarly, aggregates operate to create the sixth sense of mind and the mind-objects that appear in the dimension of time. Now we have a complete picture of how the sense of presence, the I, appears and willfully creates karma.

What we call the I is nothing more than tendencies forming aggregates or skandhas, and the mind-sense is just one of the dimensions of these skandhas. Each skandha has six dimensions, and all of them fill out physical reality. Each of the six dimensions or derivatives comes from the senses. These are touch, taste, smell, sight, sound, and mind. The most important is the mind, for in it lie the patterns that focus all of the rest. The mind exists everywhere, yet what you feel to be your solitary mind is an illusion no more real than a drop of water claiming uniqueness from the ocean that lies before it. Even if that drop evaporates, it never loses its essential quality of water. Even though that drop fills a cup, is boiled to steam, lies at the bottom of an ocean, or appears in the burning of liquid oxygen and hydrogen in the exhaust of the latest spacecraft mission, it still has its water essence. This essential water-ness is the reflection of the water-soul. And your soul is no different.

This is hard to get used to for I feel so special. The hard part is to get over this specialness and realize you are special because you are everywhere in each and every sentient life-form you meet, have met, will meet, and have never met or even never will meet. The pain you feel, the sorrow you experience, the joys of your life, are all no different from any other human being's. The sense of I-ness you feel right now is the same sense of I-ness felt by each being. That sense, changing yet unchanging, is the formation of mind-objects created by combinations of these five skandhas with their associated five spatial derivatives and one temporal derivative in all of the many forms possible—essentially 5 billion of them today.

This may certainly seem strange. We are facing a population explosion. Predictions indicate that without intelligent management of human re-sources, there may be over 10 billion *souls* on earth by 2150 or sooner. Where do all of those souls come from? Are they really separate souls and should they be given every human right, when it may turn out that the earth simply can-not sustain that number?

Consider, before you answer this, that from a Buddhist viewpoint there is no soul. From the perspective I espouse here, there is soul. However, we may be saying the same thing. What the Buddha means by no-soul is exactly what I believe is the one and only soul that exists. All others are literally skandha-based reflections of that one and only.

One (the personal sense of being and that sense you feel inside of you) is always present in all life-forms. That **one** has been, will always be, the one and only. Stand up and take a bow. The **one** might wonder: Will I reincarnate? The answer **one** gives is: Most definitely, for **I** have never died nor was **I** ever born.

In the next chapter, I'll show you how all of those 5 billion souls "out there" now and even those 5 billion yet-to-come are exactly you. In case you are wondering about all the ones that have departed, no worry. They are still around. They are, because you are, aren't you?

CHAPTER 12

How Many Souls Are There?

Sentient beings are numberless;
I vow to save them all.
Delusions are inexhaustible;
I vow to end them all.
The gates of the Dharma are manifold;
I vow to enter them all.
The Buddha way is supreme;
I vow to complete it.

—The Bodhisattva Vow[1]

HERE WE SHALL SEE, from a quantum physics perspective, why there is only one soul and how it is an illusion that many souls appear to exist. In chapter 11 we considered how skandhas (aggregates of tendencies to come into being) produce the semblance of a person, a self, or an incarnated soul. The feats of soul-incarnation, reincarnation, and karma—the willful actions taken by the soul guiding the body—are fundamental, arising from the vacuum as assuredly as the big bang and persisting as surely as the laws of physics. Since I have described these activities of the vacuum as following laws of quantum physics, we might get the impression that the soul is merely a convenient label and doesn't exist in and of itself.

However, what I have tried to establish so far is the need for a new vision of the soul as a process; something that exists beyond the confines of the flesh, yet is intimately tied to the bodily boundaries—a new physics of the soul, if you like. Ancient Greek philosophers had difficulty with the soul, imagining it to be made of the finest material. The old physics tried to put the soul into mechanical materiality. Since the soul was not material, the old physics found it did not exist and the effort failed. Adding in the modern knowledge of massless electromagnetic fields doesn't help. The electromagnetic energy image still qualifies as an old physics representation of the soul.

With the new physics-soul we deal with a massless field of conscious potentiality (an "entity" that is both not real in the materialistic sense and, since it can change the behavior of bodily matter, not unreal either) that changes and alters each person's reality. Even calling it a process restricts it too much. Perhaps this is too confusing. Perhaps we should rid ourselves of the concept of the soul, as the Buddha did. But I believe this would be a serious mistake. It would be tantamount to attempting to redefine physics purely with mathematically-real quantities—a dream some physicists still hold to—in spite of the overwhelming evidence of the imaginal component required to formulate the laws of physics correctly and to predict and understand the properties of matter.[2]

In a similar light, the soul is imaginal and does not arise as an epiphenomenon from complex interactions in matter. If one bases the assumptions about the soul on its arising from matter, although it leads to some progress regarding the mechanisms of neural communication, it also leads, as I have been stating all along, to a blind alley. It answers the wrong question.

But there are right questions to ask. Is the Buddhist notion of no-soul the same thing as the quantum physics notion of soul? Can we even count the soul as we count ordinary objects? If the soul is a process existing outside of the physical domain, how should we count it? It is very difficult to understand the mathematical thinking one needs to do when dealing with such things as uncountable quantum waves of possibility. Such things as counting, a normal operation when dealing with apples and oranges, play no obvious role when dealing with seemingly immaterial objects like the soul and, as we shall see here, the mind.

This is what concerns us here—the answer to the question of the countability of the soul. The notion of there being a large and perhaps infinite number of independent souls has far-reaching consequences for all humans. Many religions take it for granted that there are a large number of independent souls. Some even accept that there are additional souls in heaven waiting their turn, like so many shoppers in a queue, to incarnate. I hope to prove to you that this is not the case. All we need is one.

The One and Only Has Many Faces

Is there really only one soul? It seems not so as I gaze at the world outside my window. Living in a metropolis has its advantages, yet sometimes, like Greta Garbo, "I want to be alone." As I struggle through sidewalk traffic jams outside

Macy's windows at Christmastime or attempt to get across the Golden Gate Bridge at rush hour, I am faced with crowds of people. Strangers gaze back at me, none of them seeming particularly friendly or even unfriendly. They are just there, taking up space and slowing me down as I rush through time. But are they? Or is this another illusion, as the Buddha might put it? Are those strangers really so strange? Do they have souls, like me? Or are they all elaboration of a dark feverish mind's fantasy? Do these countless souls really take up space and move through time?

There are times, though not too often, when I sense those strangers really aren't that strange, but are somehow connected to me. Sometimes this feeling happens to me after I have lectured to a large group of people and someone comes up to me to exchange ideas or experiences, or to tell me a story. I get a sudden flash of recognition, a sense that I have either been here before with this person or that somehow, I have lived or experienced what the person has gone through.

Is anyone else out there besides me? At many times through my life I saw myself in others and considered that my uniqueness is the same as that of all others who experience life. Although life is sometimes frustrating and often predictable, I don't regard it as frivolous or just as mechanical motion. I know since I feel and suffer my own existence, then so do you. I must then value your life as much as mine—an instinct that gives rise to civilization.

Once, in Jerusalem, after a particularly tiring day touring the old city and the only standing bulwark of Solomon's temple, I came across a stranger as I left the gated hilltop walled Mecca. I first saw him from a distance. As he approached I noticed he was old, much older than I was then, but, just about my age now as I recall this. He was wearing a ragged, torn black coat. He had a long gray beard and appeared to be very poor, a beggar, I surmised. Yet he did not beg or bother anyone as he made his way along the road just outside the enclosed city. We were on opposite sides of the road, so his path was directed towards me into the city as I made my way back to my apartment near the Hebrew University.

I felt a sudden rush of caring. I felt tears welling up in me for no reason. I rushed up to him and gave him several Israeli pounds. He took them, looking at me with no expression in his face, not saying a word. It was as if something was coming from the dim past, something that had lasted and remained in the air for a long time, like a non-verbal exchange that had to be made, some kind of ancient debt that had to be paid, or some kindness that in spite of my normal selfishness had to erupt. It was wordless, soul-to-soul talk. He put the money in his pocket and continued on his path. I turned and watched him make his way. Not once did he stop or turn back or ask anyone for anything as he continued. I watched him until he became a small black dot, and then he disappeared through an arch in the wall, into the mystery that has always been ancient Jerusalem.

In some intuitive manner, something inside me and something inside him resonated. Perhaps, although our economic conditions bore no obvious similarity, I had seen myself in the future and he had seen himself in the past.

Let Me Count Thy Ways

Consider what it would be like for you to meet up with your perfect double. The problem is, which is the real you? The solution is, each of you has perfect right to that claim! In a *Star Trek: The Next Generation* episode, one of the main characters, Riker, encounters his double on a visit to a mission site (the result of a transporter accident several years earlier). It seems Riker-2 is still in love with Riker's former love interest, Troi, the psychic/psychologist on board, and therein lies a good yarn. How are the two Rikers going to get along?

Well, I won't tell you what happens in the story, but I would like you to consider what it would be like from the double's point of view upon meeting you. In fact, if you think about it at all, you'll find yourself in a real identity crises.

We are interested in the old feeling of comfort we call living in a body. We sense we are uniquely located, and to be elsewhere seems unthinkable. I suggest the unthinkability of this situation exists in both yourself and your double, assuming he or she is a perfect double.

Each of you feels he or she is the only one that should be around, and the other is somehow an impostor. The impossible answer is that each Riker feels he is the one and only real Riker. That feeling of subjective awareness remains in each, and as each Riker has essentially the same genetic disposition, each feels it identically.

Doubles, Triples, Quadruples, How Many Am I?

Sound like woo-woo to you? In looking at the role played by conscious beings in the universe of facts, one is usually faced with a multiplicity of awareness. In the following, we see how three people separated at birth nevertheless enjoyed the same adventures and, at times, even the same thoughts. How can one explain this? One explanation states that knowledge enters the mind of an observer beyond space and time limitations. This turns out to be a big hint about how the soul operates.

This real-life example was reported by author Ian Wilson in his book *The After Death Experience.*[3] Three American baby boys, born in 1961 as triplets, were separated at birth from one another and their mother and put up for adoption. Each grew up totally unknown to the others. During 1977 and 1978 all three underwent psychiatric treatments, quite independently of one another and for nearly the same reasons. Each felt something was missing from his life and was uneasy not knowing what it was. According to Wilson, neither they nor their adoptive families had ever been told they were separated triplets. Since even the psychiatrists had not known this, they put each triplet's malaise under the banner of "adoption syndrome."

Nothing would have ensued if one of the triplets, Bobby,[4] hadn't, by sheer chance, attended college in New York state. Quite frequently, Bobby was accosted by other students, strangers to him, who insisted that they knew him as Eddy. When Bobby clarified the mistaken identity, the other students would swear that Bobby looked "exactly like" Eddy. Eddy, it turned out, had attended the same college the year before. Finally one of Eddy's friends, having met Bobby several times at school, arranged for the two of them to get together.

Upon meeting Eddy, Bobby went through a strange experience:[5]

> It seemed forever. . . . After that, I said, "Oh, my God"—and simultaneously saw myself saying, "Oh my God." I scratched my head—and saw myself scratching my head. I turned away, and saw myself turning away. Everything in unison, as though professional mimes were doing this.

Finding each other and telling their stories led to newspaper and other media accounts. Later the two met their remaining brother, David, after David had read about the pair in the newspaper. When all three compared notes, they found they had a lot in common. Each liked Italian food, each smoked the same brand of cigarettes, and each desired the company of older women. Each, on several occasions, had dreamed of having an identical-looking brother, and each had suffered the same sense of loss of the others.

This is beginning to sound like *The Twilight Zone.* Can anything in the world of modern science explain this strange connection? If we enter into the deepest philosophical implications of quantum mechanics, deeper than Plato's cave ever was, we find some gold. If I base my argument upon quantum physics, it turns out that the entering of knowledge into the mind of an observer cannot be localized at a specific point! Thus, when you learn something, in spite of your sense that your mind lies within your skull, according to this argument, it does no such thing. The pattern of that learning in the mind, if I may think of learning as a pattern, has no space or time coordinates.

To put it in another way, the mind of an observer—the field of consciousness—exists outside of the confines of any brain. I see this field as a purely imaginal concept, although it is not imaginary. While the triplets were physically separated and countable as three beings, their minds and their souls were, at times, one. During these times, each action of a single triplet had simultaneous meaning for all three. Each action entered a single consciousness and at the same time physically manifested in all three. What does that say about the minds of the triplets? Are they connected? Yes, so much so that during these times, they had one mind, one consciousness, and, according to the argument presented shortly, so do we all.

Why only one consciousness? First of all, without this solitary consciousness and without the perceptual action entering into that consciousness, no physical phenomenon occurs. This refers to the peculiar property of the quantum physics of consciousness. No matter how we shake and dance and maybe wish this fact would go away,[6] the pesky truth remains that events occur in consciousness whenever an observer carries out any measurement of a quantum-mechanical system. This means facts are not facts until they are observed facts. Falling trees in forests do not make sounds unless there are listening and hearing ears to behold those twig-snapping, bark-rasping, and trunk-hitting-the-ground crunches.

Your Soul: The One and Only?

Granted that an event is not an event until it is a perceived event, why can't there be more than one consciousness, that is, more than one mind? Even the triplets mentioned above had, during other times, separate thoughts. Certainly my mind seems quite distinct from yours most of the time, if not nearly all the time. However, according to an argument based on quantum physics this is impossible.

The existence of a separate consciousness or separate minds leads to an absurdity, a contradiction in reasoning. Although this sounds incredible, a metaphysical physicist has offered a proof that the existence of only one mind is an experimental fact. This idea, first conjectured in the musings of Vedanta, is that even though there are many observers, there is only *one mind* or *one consciousness*. Thus, in complete accordance with Vedantic thought, two independent minds cannot exist anywhere in the universe![7] This revelation leads to another strange conclusion. The ego, or small I, the one we normally say is "me," is, therefore, an illusion, just as the Buddha said. The large I, the big I or one soul, is, however, real.[8]

The proof shows that without the action of a single consciousness, that is, the presence of one soul, the world would be continually splitting into unconscious and, therefore, unrealized possibilities rather than forming itself as it does into one conscious reality with real events appearing to that one mind as "out there." Without this single-minded soul-action in nature all other beings would exist as nothing but mindless robots.

However, my experience tells me that events are, as far as I can tell, real when they happen to me and I am aware of them. Of events happening to others, I have no immediate awareness. Thus, when I am aware of something, it is uniquely mine. As true as this seems, if my mind were distinct from all others, this experience could not occur, according to the following proof. It also follows that if my mind is distinct from others, we are all deranged robots, each thinking only he or she is conscious and all the rest are not. If we are not mindless robots, living in multiple but mindless realities, then, it will turn out, there can be only a single mind in the whole universe. Essentially this is the result of the argument.

An Absurdity of Logic

We could view this proof as the quantum-mechanical demonstration of the existence of one soul.[9] Let's see how this proof develops. The general idea for making such a proof is called a reductio ad absurdum,[10] which means showing that the opposite of the conclusion of an argument leads to a contradiction in one or more of the statements of the argument's premise, that is, a reduction to absurdity. If the opposing conclusion does not lead to an absurdity, the original argument is flawed and therefore false. Usually this means one or more of the premises are false. For example, suppose I assume:

1. To any given number greater than zero you can always add another number greater than zero and the result is a number greater than either of the previous numbers.

Seems reasonable. Take 150. It is greater than 0. Next take 7, also bigger than 0. Add them together and you get 157, certainly greater than 7 and greater than 150. Next I conclude:

2. There is a final number, Ω (omega), and no number can be greater than it.

So far there seems to be no problem. But, the absurdity immediately arises when I add one to the "final" number Ω, giving $\Omega + 1$. If this new number is greater than the "final" number, it violates 2. If it is not greater than the "final" number, it violates 1.[11] Hence, reductio ad absurdum.

Bass's Proof

To show the unity of the soul, which I shall take as equal to the seat of consciousness or mind, I shall follow the presentation[12] of the argument as first shown by physicist and philosopher Ludvik Bass. Bass was a student of the great Austrian physicist Erwin Schrödinger, who won the Nobel Prize, discovered quantum wave mechanics, and was a Vedantic scholar. Bass based his presentation on Schrödinger's early writings.[13] Assume the following premises:

1. The human body and its central nervous system function in complete accord with the laws of nature as we understand them. This includes the laws of quantum mechanics.

2. I am aware of knowledge entering my consciousness.

Statement 1 certainly seems reasonable considering everything we know about the manner in which bodies and nervous systems function. Statement 2 also seems obvious. Of course, as you read these statements to yourself you should assume the "I" in the statement refers to you. Your experience tells you when you learn something new you are certainly aware of the occurrence. As far as you know, when you learn something the experience is uniquely yours. Other people around you may or may not learn something new when you do. Your "I" is tacitly assumed to be independent from all other "I"s in the world. You have the right to assume your consciousness, your awareness of the world around you, is distinct and independent from anyone else's awareness of the world around her or him. In other words, it would seem that 1 and 2 go hand in hand with a third assumption:

3. There exist at least two independent conscious minds in the universe.

Of course, if 3 is true, it would follow that there are certainly more than two independent minds; indeed, there are as many as one would need to fill in the usual picture of human society, today about 5.5 billion.

These three premises seem to be perfectly consistent with each other. But, like my simple Ω-number example above, they are not. It turns out if 1 and 2 are valid, 3 must lead to a contradiction in 1, 2, or both. In other words, the opposite conclusion—*There does not exist at least two independent conscious minds in the universe, that is, there must be only one mind*— does not lead to a contradiction in 1 or 2. Hence this opposite conclusion must be correct. To see how this arises, consider this simple experiment:

A robot, a sophisticated, independently observing system called Robotina, sits in a room watching a tiny spinning particle. Robotina uses a complex amplification system containing a magnetic field. She is an android robot with human features and, like HAL in Kubrick's *2001*, has nearly human capabilities. Nevertheless, she has a computerized simulated nervous system that follows the laws of physics as we understand them, including, of course, quantum mechanics. Accordingly, Robotina's nervous system, being a simulacrum of your own and everyone else's, follows assumption 1. Thus when the spinning particle is observed by Robotina, those observations follow the laws of physics, including quantum mechanics, when they register in the robot's memory.

Now, in chapter 7 I explained how strangely spinning particles behave because they follow quantum mechanical laws when they interact with other physical systems. Specifically, the direction of a particle's spin axis can only be observed to be *up* or *down* with respect to any direction that Robotina arbitrarily *wishes*[14] to set the detection machine's magnetic field. So it would appear to follow that Robotina can only see *spin-up* **or** *spin-down*. The **or** is emphasized to indicate that this is a computerized mechanical mathematical operation which in specifying one of the observational choices, the other is completely annihilated and cannot be chosen. If Robotina sees spin-up, her nervous system will reverberate accordingly and the state of Robotina's memory or mind will be, "I see (register in computer memory) the particle has spin-up." We will call this memory-state Robotina-up. If, on the other hand, Robotina had observed the spin pointing down, we call that state Robotina-down, and Robotina's nervous system would reverberate accordingly.

But remember, in quantum mechanics things aren't quite this simple. Robotina's nervous system must follow the rule of assumption 1. This means that, according to the laws of quantum mechanics, after Robotina makes *her* observation, both Robotina and the spinning particle enter the "twilight zone" of what I shall call the shady gray *both/and* worlds of possibilities rather than the definitive black or white *either/or* world of actuality.

Let me explain. According to assumption 1 and the laws of quantum mechanics the state of Robotina and the particle will be both Robotina-up and spin-up **and** Robotina-down and spin-down. Again I emphasize the **and** to signify that it is also a computerized mechanical mathematical operation which specifies that neither one of the observational choices or the other is observed, but that both must be observed as a single overlapping vision (see figure 12.1).

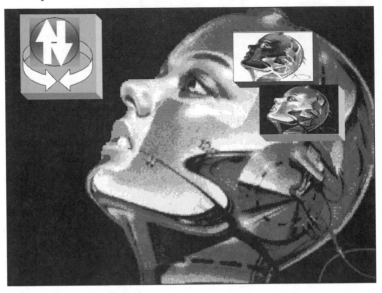

Figure 12.1. Robotina's Schizophrenic World. Robotina, having a computer memory that follows the laws of quantum physics, sees the spinning particle with both possibilities simultaneously. Her "mind" contains an overlap of both possibilities together with her memory of herself (shown here separated in memory) having each possibility.

You can read this as Robotina's *memory* containing *conditional* knowledge: If spin-up, then Robotina-up **and** if spin-down, then Robotina-down.[15] Both possibilities of concordance must co-exist in Robotina's nervous system. In other words, the state of the universe, including Robotina's brain after Robotina's measurement, is a conditional one and not a definitively unconditional world where Robotina sees either the spin is up or the spin is down.

Now you come upon the scene. You ask Robotina what she has observed. She gives you an answer, possibly in a computerized voice, and that response enters your independent mind as a fact conforming to assumption 2. Accordingly, the knowledge in your mind regarding Robotina and the spin of the particle will be either "I know that Robotina-up and spin-up" **or** "I know that Robotina-down and spin-down." You will never hear both answers when Robotina talks to you. You will not ever know both, only the single case of one **or** the other. What was a conditional, uncertain possibility has for you become an unconditional, certain fact. For now you do know (unconditionally) what Robotina previously knew (conditionally) and you now know what the particle is doing (unconditionally). As far as you are concerned

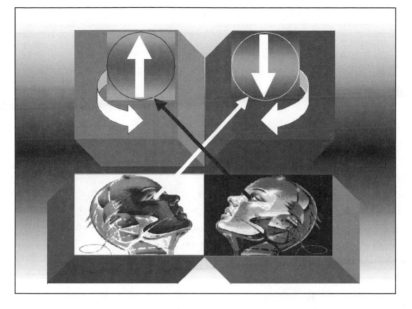

Figure 12.2. The Stopping Buck of Observation. After you inquire, the state of the particle is either one way or the other, Robotina has told you a single answer; the particle is either spin-up or spin-down, and you know unconditionally the way it is!

Robotina has responded to you in a perfectly human way. She has answered your question in a completely consistent manner. She has behaved as if she knew the answer all along and made up her mind well before you asked.

But before you came along, you knew, according to the rules of quantum mechanics, that Robotina was not of a single mind! Once you asked her what was going on, her mind, as well as the spin of the particle, immediately jumped from a conditional state into a single unconditioned state. You, the ultimate observer, the creator not only of your own experience of knowledge entering your consciousness, but also the provider of the clarifying action of resolving Robotina's "mental anguish," can take all credit! Your knowledge is private and your single, independent, unconditioned mind is clearly distinct from Robotina's prior conditioned robotic mind. Your present mind is in a very different state from Robotina's prior mind. You know it's an up or down world. And before you asked, Robotina, following the laws of nature as in assumption 1, had to have her mind split into two simultaneously occurring possibilities, both Robotina-up and spin-up **and** Robotina-down and spin-down. Thus Robotina is a nonconscious[16] robot unable to make up her mind, simply because she can't have such an experience as making up one's mind.

Paradoxically, Robotina, being a robot, didn't know, until you came along and resolved the difficulty by telling her so, that she lived in two worlds. You could do this because you, a conscious being according to 2, know only one of those worlds, never both. In fact, if you think about it, you have never experienced anything like a superposition of both worlds at the same time. You always have a distinct and separate consciousness that follows 2. It is distinct from Robotina's nonconsciousness that apparently follows 1 and, therefore, you have a distinct and separate consciousness from all robots who, as you can see, are acting strictly according to 1. You could argue that the role of your consciousness is to follow rule 2 and thereby to resolve all unaware conflicts. All robotic overlays of unaware possibilities pop into conscious realities only when your consciousness acts. When all other robot minds act, when they take in knowledge, it is nonconscious data entry entirely and these robots are subject to rule 1 only. They are and they remain in the imaginal realm of **and** unless you perform a conscious act that separates what they hold in memory into the world of **or**.

Suppose that Robotina was not a robot. Suppose, like the straw man in *The Wizard of Oz* or the character Pinocchio, Robotina becomes real, Robotina becomes a Roberta. Now everything we have said about Robotina clearly becomes paradoxical and, to some extent, nonsense when we replace it by Roberta (a real person). For example, you could ask Roberta, after you ask her what she saw, if she knew what state the particle was in before you asked and most assuredly she will say yes. She would say she knew what the particle was doing independently of your asking. She would say her world was clearly one way or the other; the particle had its spin up or down. Roberta won't appear to herself to be a schizophrenic, unthinking robot. But, in spite of Roberta's denial, until you ask her what she saw, you would have to conclude that since Roberta's nervous system is a physical apparatus following 1, in spite of any protests she might make, her mind registering the contents of that nervous system must be in the **and** state.[17] In other words, as far as you are concerned, Roberta is a Robotina—a machine recording both possibilities.

Here Bass argues this result is unacceptable. The problem seems to arise the minute we bring in the existence of a mind acting independently from yours. How can the action of Roberta's mind follow the laws of awareness given by 2, as she would put it, and the laws of nature as given by 1, as you would put it, simultaneously? How can Roberta's mind following 1 be in the **and** world while Roberta, following 2, professes no such duplication even before you asked?

Following 3, Roberta must have a mind different from yours, but that leads to this contradiction. From your point of view, entering data about the spin of the particle into Roberta's separate and apparently schizophrenic mind followed assumption 1; this led to a conditional world of **and**. Even though you might be tempted to accept Roberta as a real person, her nervous system had to be totally in accordance with 1. In effect, Roberta had the robotic mind of Robotina.

Schizophrenia Denied

Some physicists tend to regard the above paradox as quantum mechanics' inadequacy to take the nervous system into account. In effect, they deny assumption 1. They instead assume when a fact of knowledge enters and modifies a nervous system, quantum physics, usually considered to apply only to microscopic phenomena, no longer applies. However, so far there is no evidence to confine the validity of quantum mechanics to microscopic atomic phenomena[18] nor any evidence the brain follows any laws other than natural ones, including quantum physics laws.

We might question assumption 2 except it is a very strong assumption seeming to have no contradictions. Descartes said it all: *cogito ergo sum* (I think therefore I am). Or, if you will, the mind-sense is very strong, perhaps even stronger than the physical senses that lead to assumption 1. You must assert to yourself at least that you know what you know directly without having to read the mind of anyone else. In fact, if we take 2 as a statement regarding the action of consciousness upon any physical system, namely the conversion of the plural **and** world to a single **or** world, we might change Descartes's statement to *opto ergo sum* (I choose, therefore I am).[19]

There must be contradiction, a war of worlds, between one kind of awareness (**or**, meaning one or the other) and the other (**and**, meaning one and the other together). Could this merely be semantics—not a war of worlds but a tempest in a teapot of words? Going back to the robot Robotina, suppose that Robotina's brain-state after the observation of the spin was **or** and not **and**. Will the physical material consequences be any different? The answer is yes. Experiments can be performed that would detect the presence of both **and** parts of Robotina's nervous system coupled conditionally to the particle's spin that would yield different results from the case where Robotina's mind and the particle were either one **or** the other. These would be, undoubtedly, extremely difficult experiments to carry out, perhaps even impractical but, nevertheless, would provide very differing physical results.[20]

Not convinced? One could even imagine there being a third separate mind, say, your professor of science who, unbeknownst to you and Roberta, was performing an experiment on both of you. She would claim, being in a position of authority, both you and Roberta were not the ultimate minds to consider the effect of the spin-measurement, that you and Roberta were both in nebulous worlds of conditional possibility and not in the distinct world of unconditioned fact that she was in as a result of her knowledge entering her consciousness. She would claim that while you, unknown to you all, were in schizophrenic multiminded illusion, in spite of your protests, she was not.

Thus we could imagine a nested set of observers going all the way up to some ultimate observer who would insist, being the final link in the chain where the buck must stop, that she was the ultimate choosing observer and that you all, including the professor, were schizophrenic "conditionals" until she came on to the scene, unconditionally.

What is wrong with this scene? The paradox arises as soon as we ascribe a separate consciousness to another person, Roberta, for example, positing a mind independent of yours such that, correspondingly, events in her consciousness must be independent of events in yours. Assume Roberta is a Robotina and the problem vanishes. The nonconscious world, nonminding its own business, can certainly be independent of the conscious world. But once consciousness occurs, the bi-location of that consciousness becomes a problem leading to paradoxes.

Therefore, the problem, according to Bass, is assumption 3. With assumption 3 in effect, Roberta's (not Robotina's) observation leads to a contradiction: She says her acquiring knowledge followed assumption 2 and your independent mind says this acquisition followed 1, and, according to the laws of physical reality, it can't be both. Hence a reductio ad absurdum.

While the brain and the body with its nervous system must be conditional, subject to the laws of physics according to 1, the mind or consciousness must, by definition according to 2, be unconditional. Unconditionality arises from the action of the mind. We might put it this way: the body, brain, nervous system, and all material objects follow 1, but the mind follows 2. Brains can be numberless and nonconsciously numb, that is, without consciousness, while the mind must be unitary and conscious!

As one wise person sagely put it[21] (and I want you to consider both meanings of the following questions) . . .

What Is Matter? Never Mind!
What Is Mind? No Matter!

The double-meaning double question and answer leads to physical brains and all matter nonconsciously existing in the **and** plural world of quantum physics until mind—and that means mind alone—comes along, pops the question, and makes an observation. When that happens matter becomes conscious, so to speak. Where does this leave us? It doesn't matter who asks. It doesn't matter where or when the measurement, the action of mind, takes place. Knowledge's entrance into consciousness stops the buck there and then, and the plural **and** world is immediately reduced to a single **or** world. Thus, while Robotina doesn't even *mind* not having a mind, Roberta does!

Roberta's mind and your mind cannot be independent because the action of consciousness cannot be located in space or time. When the action occurs, the results of the action are, of course, located in both space and time. If there were two independent minds, then according to you, Roberta's mind and the particle would exist in a plural **and** world. Meanwhile, according to her, in contradiction, Roberta's mind and the particle would exist in a single **or** world.

In a similar manner, while your brain doesn't even mind not having a mind, your soul does! Your mind and all other minds cannot be independent. Since our knowledge and/or our experience tells us 1 and 2 are true, 3 must be false. It must follow that the entering of knowledge into consciousness always follows 2, regardless of when and where it happens. This includes Roberta's experience of the spin-knowledge entering her consciousness, even though I may not be there to act on Roberta's behalf and resolve her dichotomy! Or, in other words, for whatever reason the universe has come to be as it is, there cannot be two independent conscious minds in it. Consciousness is a singular word whose plural is unthinkable even though the illusion of independent minds exists and persists. Such is the nature of *maya*.

Maya's Veil: You Go Your Way and
I Go Mine

We are all living in an illusion. Consider that of nearly six billion independent minds, yours follows premise 2 above, and all nonconscious others follow premise 1. They exist simultaneously.[22] The amazing result of all this? When you pass from your mortal coil, we all go with you, for there will be no ears to hear, even though there are many listening, and there will be no eyes that see, even though there will be many looking. As physicist Richard Feynman, attempting to describe how awareness changed the universe when the first conscious entity appeared in evolution, put it, "a mite makes the sea roar."[23]

In other words, suppose we reject the argument put forward by Bass and take it that two or more minds can really exist, separate and unequal—one of them being the ultimate observer, namely someone we call *A*, who must follow premise 2, and all others (called *B*s) who, until *A* comes along, must, according to premise 1, remain in the nebulous world of **and**. Can we get away with this without running into any serious paradoxes?

Not quite. Consider this puzzling situation based, according to Erwin Schrödinger,[24] on Indian Samkhya philosophy. Assume two bodies, A and B, exist and A is put into a room containing a view of a garden while B is put into a dark room. Suppose next that A and B are interchanged so A is in the dark room and B is in the room with the view of the garden. The result of all of this is that first there was a view and now there is definitely no view, all is dark. How can this be? The answer is that A is the ultimate observer, you! Before you saw a view of the garden and now, as you stand in the dark, there is no view. What happens to B is of no consequence because *it* is not conscious, as you are.

Does this mean that while you stand in the dark, there is nothing to be seen everywhere in the world? So it would seem, based on your own experience. Remember, we are assuming here that your mind is ultimate and the creator of all unconditional reality while all other minds sit in the quagmire of conditioned reality.

To see how strange this is consider the following: If we now identify the totality of consciousness, the soul of the world, with only your body, we must conclude that when your body dies, the soul no longer exists. That would leave the world without you in a very sorry position of continually and robotically following 1 with no hope of ever reaching unconditioned reality. Such a world would be very strange indeed if there were any consciousness capable of seeing it.

Maya Accepted, Then What?

On the other hand, if we accept the proof that there cannot be two independent minds, we are faced with another conundrum. Why does anyone have to ask anyone else anything? Why are there questions at all? How come, if only one mind is present, there is so little telepathic communication going on? If there is any telepathy, it is surely a very weak and erratic phenomenon. Some might argue there is no telepathic communication occurring between minds at all, thus positing for 3. In effect, the veil of maya is so good that it has concealed the beauty of the soul's one-mindedness nearly completely. Can we explain this efficacy with science?

Although one-mindedness is the natural mode of consciousness, that is, telepathic communication is normal for all species including ourselves, forces of evolution make it advantageous for telepathic insulation to arise, especially in intelligent sentient life-forms like us!

J. B. S. Haldane, a noted British geneticist, once pointed out in this regard:[25]

> We should expect [telepathic] phenomena to be unusual as, from the standpoint of natural selection, a person who habitually experienced other people's sensations would be less fit than a normal person. I should not be surprised if our mental insulation turned out to be a special adaptation.

Thus we would conclude that the veil of maya is vital for the survival of the human species. Of course, we might ask if there are any species in which this mayan veil is lifted. I suppose fish, ants, bees, and birds show remarkable telepathic signaling. A flock of birds making its way across the sky at sunset exhibits this phenomenon of one-mindedness. Ants and bees seem to go about their ways in purposeful single-mindedness, always "thinking" about the hive over the individual.[26] These species clearly show us how one-mindedness produces a tribal effect, one many people today desire, with the growing isolation we feel from each other because of modern technology. By having such insulation, one could argue, one mind can have a stronger effect by acting in a more concentrated manner in the human species, namely, individually.

Consider where one-mindedness clearly is important, namely, inside an individual body. Charles Sherrington, a noted neurologist, indicated in his work that there must be a plurality of mind-entities in any one living body, each quite capable of separate cogitation. As he inquired about the mind/body:

> How far is the [one] mind a collection of quasi-independent perceptual minds integrated physically in large measure by temporal concurrence of experience?

In the caverns of your body, while the multitudinous chemical and electrical reactions occur, no insulation between the quasi-independent minds arises in all of the tasks carried out. While you are driving your car, thinking about dinner, listening to music, attempting to balance your checkbook in your mind, you are one, with one consciousness made up of all those quasi-independent minds! Thus, we might inquire how the idea of plurality arises at all. Simply, there are clear indications that the sensations I experience are localized in space and within time. There are clear indications that the thoughts I experience are also confined in the same manner, namely, to my body and to none other. Since there are a lot of bodies out there, it would appear that each of them must have similar privacy within their caverns.

Yet we know that in spite of all of this plurality we experience around ourselves, none of it can be explained scientifically without using quantum physics. And when we bring the ubiquitous quantum into the fray, much of the insulation breaks down, not only within the confines of a single body but also between any two interacting bodies.

For example, in the above exercise of Roberta and the particle, suppose we replace Roberta with another spinning particle, so after the two particles interact, one of them flies off in a box to the moon and the other remains in another box on Earth. Now we place astronaut Armstrong back on the moon and ask him to hold his box and open it at a precise time. According to 1 the particles are in the nebulous world of **and**. That means the particle on the moon, like Roberta's brain, is conditionally connected to the particle back on Earth. Let's label the particles and their respective spin possibilities according to where they are and in which direction their spins point: moon-up, moon-down, Earth-up, and Earth-down. Accordingly their state is: both moon-up and Earth-up **and** moon-down and Earth-down.

Now you come along and open the Earthbound box at precisely the time Armstrong opens his. You see either Earth-up **or** Earth-down. You send a message to Armstrong, telling him, "When you opened your box, you saw what I saw; either you saw moon-up or you saw moon-down, exactly as I did." This means that you are in the up world and you send that message to Armstrong. Or you are in the down world and you send the corresponding message. You make the prediction because consciousness has acted and knowledge has entered. It doesn't matter where or when it acted. The effect it had in this case was simultaneous in two places, separated by nearly three hundred thousand miles. As far as consciousness is concerned, it could have been a million miles or clear across the galaxy. Consciousness has no unique location. Your mind is the only mind there is, the Mind of God!

We could conclude from this that nature has her reasons for erecting the veil of maya. Natural selection draws the veil across the face of God so we can develop as individuals. We each have little telepathy in order to survive. In nonthreatening situations, less telepathy implies greater survival for the human species. Normally we need our illusions in order to survive.

But things are changing today. Survival of our species appears to depend on our spiritual revival. We need to re-invoke the sacred in all of our lives. This means the illusionary curtain that separates the one mind into the countless quasi-independent minds of humanity may be fraying at the edges. To unveil the curtain, we need to listen to our soul and intuitively sense its presence and needs.

CHAPTER 13

Soul-Talk

The soul has the structure of a point in actuality . . . , and the figure of a circle in potentiality. It pours itself forth from that punctiform abode into a circle. Whether it is obliged to perceive external things that surround it or whether it must govern the body . . . , the soul itself is hidden within . . . It goes out, then, to the exterior of the body according to the same laws by which the surrounding lights of the firmament come in towards the soul that resides in a point.[1]

—Johannes Kepler, 1619, *Harmonices Mundi*

If the existence of the Soul is admitted on the basis of the argument that it is self-luminous, that knowledge, existence, and blessedness are its essence, it naturally follows from this that . . . there was never a time when It did not exist; because if the Soul did not exist, where was time? Time is in the Soul; when the Soul reflects Its powers on the mind and the mind thinks, then time appears. When there was no Soul, certainly there was no thought, and without thought there was no time.[2]

—Swami Vivekananda

Oh, Oh, I am coming,
I am coming through
Coming across the divide to you.
In this moment of unity
I am feeling ecstasy,
To be here, to be now,
at last I am free.

Yes, at last, at last
to be free of the past
and of a future which beckons me.
I am coming, I am coming,
Here I am,
Neither a woman, nor a man.
We are joined, we are one
With a human face.
I am on earth
And I am in outer space.
I am born and I am dying.
Yes at last I am free.

—"Coming," from the movie *Orlando*, by Sally Potter,
Jimmy Somerville, and David Motion;
published by Copyright Control/Virgin Music

Thus there are these two streams, one from the past and one
from future, which come together in the soul—will anyone
who observes himself deny that?[3]

—Rudolph Steiner, German anthroposophist, 1861–1925

THE EXISTENCE OF A SELF and a separate soul suggests that the two are not identical, which is probably the most important theme of this book. Here we are going to present a model of how soul speaks to each of us through its many illusions called selves. With the insight gained in the previous chapter, we concluded that consciousness—the field by which the soul communicates with the self—is a singular noun whose plural is meaningless. Yet you and I each feel we are uniquely conscious. Certainly, at times, we have heard a nagging voice inside our heads urging us to take some form of action or possibly to cool it. Even if we failed to hear that voice inside our heads speaking as if someone else were actually there, we are aware of being conscious. Much as if someone had turned on a television receiver when no station was broadcasting, the buzz of our conscious minds, like a hive of bees in a field of clover, asserts itself. If this buzzing field is directly attributed to the presence of the soul within each of us—and, in the previous chapter we saw that this field cannot be personalized based on a reductio ad absurdum—you and I face the paradox of a single potential voice, a world-soul residing in each of us, capable of speaking to us as if broadcast from a single television antenna, no matter where we are in space or when we exist in time.

In chapter 12, in the example of Roberta and Armstrong on the moon, we considered the effects of this single consciousness—the soul-voice. These were instantaneous, even though the recipients of that effect were far apart. Therefore, we conclude that this soul-voice clearly exists not only within each of us, but also without. I am convinced this voice can be heard inside of you, and I assume that you are convinced it can be heard inside of me. We seem to face the nonspatial or nonlocal characteristic of the soul,[4] namely, that it is in instantaneous (beyond space and time) communication with the self or body-mind.

In this chapter I want to examine further the non*sensical* (no pun intended) characteristics of the soul's communication with each of us. Communication between soul and self is difficult at best. Often the soul is not heard or becomes devastated in its attempts to reach the deeply embodied and preoccupied self. You can see an example of this in Wim Wenders's film *Wings of Desire*, in which three angels come to postwar Berlin to listen to and give comfort to Berliners. Their presence is hardly felt by adults but easily witnessed by children who see and hear them quite easily. The adults are too wrapped up in "lost soul" thoughts of despair. The souls of the Berliners are too deeply embedded in their bodies to even notice the angels' presence. The angles have come to revitalize and free the souls of the preoccupied city dwellers.

How can the soul become devastated? In chapter 3 we saw how Plato imagined the despoilment of the soul simply by its being in the body. Consider, now that we have looked at a quantum physics model of the distinction between self and soul in chapter 7, how such despoilment could take place by mere communication between the two. While the following example may differ significantly from any that Plato would imagine in his proof of immortality, I am sure you will get the point of it and possibly even see how Plato, had he thought of it, might have used it.

A Ship of a Fool

You suddenly find yourself as a passenger on a ship moving across the ocean.[5] You are in the ship's bowels and it is pitch black. You are able to move around the inside the ship, but since you don't know where everything is and because of the darkness, you bump into things. Eventually, as time passes, you learn how to fuel the ship, to keep it maintained and running, and even how to control the steering mechanism, but since there are no portholes to look out, you cannot see in which direction to go. Besides, once you get used to the darkness, there is a lot to amuse you inside the ship: TV, movies, and good things to eat, taste, and sense. But, you still cannot see the outside, so you find it pointless to attempt steering.

The ship, moving across a vast ocean, appears to be drifting without direction. Once in a while you are buffeted about as the ship seems to take a different direction. You wonder why it takes such a meandering path, sometimes going this way and sometimes in just the opposite direction. At times, particularly when the ship's direction changes, you hear a vague, almost imperceptible voice coming from outside the ship. It is calling to you. You have heard the voice from outside before but have ignored it, thinking it was only your imagination. After a while you hardly hear the voice at all.

Then, one day something goes terribly wrong. The ship appears to be jostling about, rolling, pitching, and yawing violently. You feel lost and ill to the point of terror. You begin to cry out, seeking help, crying for a vision and healing. You curse the powers that have placed you in this predicament and now you insist on knowing where the ship is heading.

The Inner Self and Outer Soul

Meanwhile, on the deck of the ship another being exists. She hears your voice, and she answers as loudly as she can. But she senses her answer is not heard. She moves closer to the hull and uses all of her powers to communicate with you. She attempts to open a portal in the hull. But the sense data she receives are overwhelming; the sounds, smells, and flavors overcome her, intoxicate her. She braves her dizziness and moves closer, yelling to you. She shouts, "Listen to me. I am on the deck, I can see, I know where you are heading, I can tell you in which direction to steer the ship if you will only listen to me."

For a brief moment you hear her muffled cry; you barely hear her instructions and attempt to listen more closely. Finally, you hear her clearly, and you are able to pay attention, steering the ship in the right direction. Eventually, after many diligent years, you make it to safe haven.

The story is a metaphor of your life. You are the self inside the ship and your soul is the person outside. This story illustrates the process of soul-recognition and the difficulty of soul-to-self communication. It indicates that the soul and the self are locked together in a duality, each somewhat helpless, without the assistance of the other, to control the ship, which is your own body, moving through life.

The Soul *and the* Self *Are Not Identical*

The story illustrates something we intuitively sense about the soul, namely, that it is somehow outside of the body and inside the body at the same time. It suggests that the soul and the self are not identical and implies the self is strongly embedded in the body, perhaps even identified with the body or body and mind, while the soul remains in some way aloof from such material concerns, but in danger of losing its ability to guide by becoming embedded more deeply in the body. It also indicates the language of the soul is not a language of logic and words but instead is one that speaks through the heart and intuition, often most loudly when we are in the deepest trouble.

The self, unable to see the outside, is lost without the direction of the visionary soul. What could the soul be saying to us and why is soul-talk so mysterious? In this chapter we find out. We see how the soul's message appears to be both visionary and compassionate, enabling the self to burst out of its egoistic shell.

Some of you may feel that you do not have a soul or, if you have, that it has never personally talked to you. After reading this chapter, you will come to a different conclusion. You all have had soul-talk experience. To understand this, however, first we need to look at how soul-talk takes place. You might call this God's technology in action. After all, when your soul talks to you, she or he doesn't exactly call you up on a telephone![6]

When the soul talks, the message is not going from somewhere else to locations in your brain.[7] Soul-talk is not only nonlocal—the strength of the message does not decrease with spatial separation from the source—it is also noncausal, meaning without prior or earlier cause. When the soul talks, you can't help but listen, and the words it speaks arise spontaneously from no location and from no earlier time. In fact, if time enters the relationship at all, it appears that the soul speaks to the self from the future. We will investigate what the soul says and whether it can be taken to be rational at all times and whether it is the soul talking or some other rambling entity.[8]

The Ancient World-Soul

According to legend, Moses went up Mount Sinai able to hear the voice of God and came down unable to hear it, with the law in his own hands. From that day forward, the Old One, as Albert Einstein called God, has remained silent. From that day forward, we have been left to our own devices and we have struggled, attempting to find life's meaning. Going back through the early history of science, in my attempt to depict this picture of how the post-Mosaic soul speaks, I discovered Johannes Kepler's ideas cited in *The Interpretation of Nature and the Psyche* written by Carl Gustav Jung and Wolfgang Pauli.

Kepler, a German astronomer and astrologer, lived from the later sixteenth to early seventeenth century. He was interested in astrology, mysticism, and the soul and its connection to the world. He was also fascinated with geometry and its relation to the sacred universe. As he put it, geometry is the archetype of the beauty of the world. In his view planets were living things and, like people, endowed with individual souls. Yet as he probed the then-emerging physical science of astronomy, he began to take a de-animated, that is, no-soul, view of nature. As in Moses' time when God became silent, the mysticism of Kepler's time was growing mute.

The Interpretation of Nature and the Psyche is two books inside one cover. The first, written by Jung, *Synchronicity: An Acausal Connecting Principle*, deals with Jung's ideas attempting to relate modern physics and unusual events, called synchronicities, together. The second, *The Influence of Archetypal Ideas on the Scientific Theories of Kepler*, is pure Wolfgang Pauli in his attempt to deal with the roots of the irrationality which, in turn, led to the growth of rationality of science.

Pauli was a well-known physicist, a Nobel Prize winner, and the discoverer of a remarkable principle of quantum mechanics, one having no connection with any previously perceived property of gross matter. It shows that the spin of all particles enables them to be deeply connected by a kind of psychic synchronistic force. Known as the Pauli Exclusion Principle, this force enables atoms of matter to be fundamentally stable and, at the same time, chemically active. I mentioned Pauli's electron exclusion briefly in chapter 6.

Pauli, a somewhat tragic figure, died in 1958 at age fifty-eight. Having had acute psychological difficulties due to his divorce in 1930, he spent some time undergoing psychoanalysis and collaborating with Jung. For the remainder of Pauli's life, they had a strong friendship, the results of which we are just now learning.[9]

What drew my attention was Pauli's realization that Kepler's life at a crucial time in the history of physical science, the very beginning of physics as we know it, spanned the twilight time of both science and mysticism. While science was growing stronger, mysticism was diminishing. Thus, Johannes Kepler is a figure worth noting for anyone interested in the soul. Prior to Kepler (1571–1630), there was no such thing as science, particularly as it is understood today with its basis in experimental investigation and mathematical quantification. Kepler was a scientist on the bridge connecting the mystical views of nature with those of the fledgling concepts of the new modern science of experimentation and mathematical theory.

Kepler is thus, as Pauli put it:

> especially suitable, since his ideas represent a remarkable intermediary stage between the earlier, magical-symbolical and the modern quantitative-mathematical descriptions of nature.[10]

Pauli may have seen himself in the figure of Kepler since he had his own battle raging between the spiritual soul and the material self in him also. Like Kepler, who had more than a passing interest in mysticism, Pauli was motivated by archetypal concepts that, while having deep roots in the human psyche, emerge as foundations for culture and discovery in nature. Thus, Pauli, like Kepler, was greatly influenced by archetypes, and if one digs deeply enough into his writings, one sees he was a number mystic.

For example, before discovering the famous Pauli Exclusion Principle,[11] he knew nothing about its ultimate connection with the physical spin and magnetic moment of the particles (in this case electrons) involved. With his discovery, emerging as pure number vision, as if it were from a deeply buried archetype or symbol, he realized the only way for atoms to exhibit their remarkable stability was for atomic electrons to possess a fourth state of existence realized by a fourth quantum number. This property, known as *spin*, and its spin quantum number have no classically accessible counterparts.[12] Somehow the spin of each electron was instantly communicated between all of the electrons in the atom enabling them to form shell-like, geometrical, stable cloud-crystals of possibility around the nucleus. Pauli found this by intuition and not physical considerations. I'll get back to Pauli's shells shortly.

The Geometry of the Soul

From the day Moses descended from Mount Sinai to the present, the ancient world-soul has remained silent. Not content with this silence and entranced with the geometry of the universe, Johannes Kepler provided the first scientific model of the soul. Kepler saw the soul as a central point pouring itself forth into a circle like radial lines connected to circular waves issuing from a stone dropped in a still pond. The soul radially moves out then, to the exterior of the body, according to the same laws by which the stars shine. Kepler believed that the individual soul reacted to specific rational and harmonious divisions of that circle.

Remember that Kepler lived during a time that spanned the ending of the dark mystical age and the beginning of the enlightened scientific era. Kepler's spiritual universe was filled with his Platonic love for order, symmetry, and beauty. In a somewhat similar way, Wolfgang Pauli's world lay, like a light-reversed negative of Kepler's picture, in the shadow of science and the sunlight of mysticism. A mystical number intuition led him to a new model of the atom. Pauli, until the end of his life, believed science went too far in its attempt to throw the psyche out of the universe. His goal was to model a science that encompassed both the soul and the physical world.

Kepler saw the planetary orbits forming shell-like spheres that fitted within structures—the regular polyhedra, all centered about the sun, quite similar to Pauli's shells.[13] Kepler believed these spherically inscribing and circumscribing regular solids were proof that God created and regulated the order of the cosmos. Pauli saw with his exclusion principle the formation of regular geometrical shells setting the order of the atom.

In his vision Kepler also saw the holy trinity[14] in the image of God, the Father, as the center of the universal sphere, Jesus, the Son, an image in the surface of that sphere, and the Holy Spirit as the relationship between the center and the surface. A movement or emanation passed from the center to the surface, representing creation. In this manner the world-soul or God communicated with His creation.

Just as the world soul communicated from the sun to its planets in the great celestial nest of inscribed and circumscribed solids and spheres, Kepler believed the soul communicated with the body by emanation: a beam of light-energy was sent from the soul at the sphere's center along radial straight lines to the surface representing the body. So taken with geometry was Kepler that he firmly believed that the individual soul had to possess the fundamental ability to react to certain harmonious proportions which corresponded to specific rational divisions of the circle. We saw this appeal to rationality in the ancient Greek Pythagorean science of music, as I described in chapter 3. This

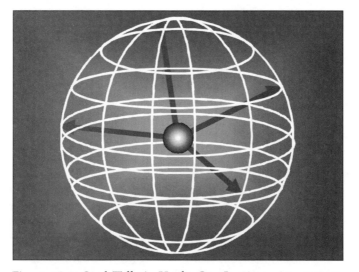

Figure 13.1. Soul-Talk As Kepler Saw It. The soul (and God), at the center of the sphere, communicates with the body (Jesus), at the surface of that sphere, by emanating radial lines of spirit.

shouldn't be surprising, since Kepler had little else on which to base a science of the heavens.

Though Kepler knew the sun was at the center of the universe, he considered astrology seriously since it was also based on rational circular measure, particularly the twelve signs and the angles subtended by those signs using the earth as the center. I don't wish to get into Kepler's astrology here, but only mention that he believed the soul flowed at birth into a pre-existent form, shaped in the earth by rays of light from the stars and planets.

In a manner similar to Moses, Kepler, and Pauli, I sense that we are in a unique period of rediscovery of the soul. As Moses manifested the law of human affairs at the cost of silencing God, and Kepler manifested the vision of a logical universe with mysticism's fading, Pauli attempted to rebuild the bridge to God's mystical universe with scientific principles. My work here, making a model of the soul based on quantum physics, follows the work of Plato, Pythagoras, Kepler, and Pauli.

Boxing the Soul

The connection of Kepler's views with a model[15] of the self and the ego, which I offered in my earlier book, *The Body Quantum*, interested me. With a slight modification, I think it useful to show how this model applies to the soul and the self. So following Kepler's lead, imagine the domain of the soul is again a volume of space. Instead of a sphere, I shall use a box to represent the domain of the soul because it makes it easier to visualize the concepts I wish to demonstrate.[16]

Just as Kepler's soul, as a point in actuality and a circle in potentiality, was able to essentially change its size by radiating light from the actual point to the potential circle, imagine that the soul is a box capable of changing its dimensions by instantaneously altering its width, thereby growing wider or narrower. I'll take it that the world-soul is represented by the wider, and the restricted or fallen soul known as the self, by the narrower. Such a sudden collapse or expansion of the box is commensurate with the notion that the soul communicates instantaneously with the self by collapsing its boundaries. Similarly the self is capable of communicating with the soul by expanding its boundaries.

Hence, in seeking each other the soul falls inward and the self expands outward. When the box collapses in width, information that was within the soul domain becomes quite localized within the body domain. If the collapsed information-waveform is undistorted, the soul *falls* into the self providing it with guidance. In this way, the self has "learned" of the soul's existence.

Within the boundaries of the box, knowledge exists in the form of standing quantum waves of possibility. We want to see how that knowledge, soul-talk, is delivered to the self. One might imagine that the box represents the whole universe or some particular region in the vacuum of space, perhaps within the space of the body.

Following Western Judeo-Christian spiritual tradition, it has long been held that humans were created in the image of God. So in what follows, we are to imagine that the world-soul, universal-soul, oversoul, or God is simply a box. I mean nothing sacrilegious by this. I am sure that God or the oversoul is not a box. In a similar manner I imagine the self as a smaller version of the soul, another box, capable of holding the image of the soul and attempting to communicate with it. I certainly know I am not a box and I know you aren't either. This is just a model representing an aspect of soul-to-body or soul-to-self communication that I'm sure hasn't been seen before.

Inside the soul-box, quantum waves of possibility vibrate up and down much like a jump rope held by two children. Such a vibration is called a standing wave simply because the wave does not propagate or move from one side of the box to the other. The pattern is quite simple, representing a bit of

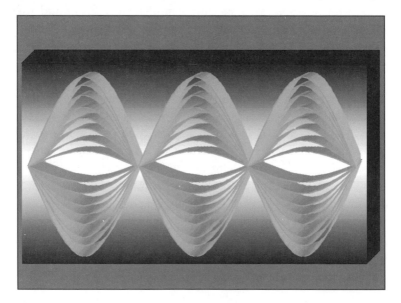

Figure 13.2. Quantum Soul-Talk (A). Standing quantum waves of possibility confined to a box. The box represents the soul and the wave its soul-talk. Here we see a wave with three maxima and two nodes. The figure shows stroboscopic instants of the vibrating wave. The pattern represents some aspect of soul knowledge. Here the self and the soul are one.

knowledge. That bit is perhaps contained in the internodal separation, the distance between two nodes, which measure one half of the wavelength of the vibration. Like any vibrational pattern, it has a frequency and wavelength and when it excites some receiver, that receiver will in turn feel the vibration. Perhaps this pattern represents a thought or a feeling, perhaps a sacred understanding of the universal vibration of *aum.*

Next the box undergoes a collapse in width. In quantum physics this is related to the so-called collapse of the wave function corresponding to the action of consciousness. When this happens, information that was spread out suddenly becomes quite localized.

Although the information is now confined to a smaller region of space, if the ratio of the widths of the box corresponds to a ratio of integers, a so-called

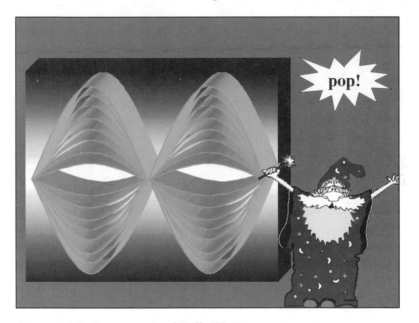

Figure 13.3. Quantum Soul-Talk (B). The standing quantum wave pattern after the box has undergone a sudden, as if by magic, "rational" collapse and the standing wave pattern is reproduced. The box is now exactly 2/3 its original length and the pattern is exactly as it was except there are only two maxima and one node. Since 2/3 is a ratio of two integers, the new length is rational when compared with the original length. The new smaller box represents the self and the rational collapse represents the soul communicating its knowledge to the self. The soul remains the same size as it was. The collapse could correspond to the original fall of the soul into the body or birth. The self and the world-soul are separate but still joined by shared information.

rational number, the pattern remains identical to the original wave.[17] In this case the information is available to this localized region—the narrower box—taken as the model of the self, or perhaps the body. The shrinkage of the box's boundaries corresponds to a form of communication of the soul to its confined space, the self. If the wave-form is identical we say that soul has fallen and has created the patterning of the self. The self has "learned" of the soul's existence. The soul and the self are vibrating together with the same harmonious pattern of aum, for example. This picture can also correspond to the soul at the moment of birth.

On the other hand, it is possible for the box to collapse irrationally, meaning the ratio of the widths cannot be expressed as the ratio of integers. For example, the ratio of the new width to the old could be $\sqrt{2/3}$. Interestingly

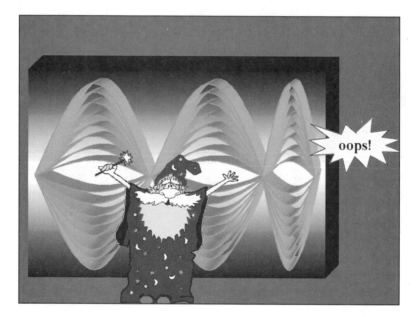

Figure 13.4. Quantum Soul-Talk (C). If the box has undergone a sudden "irrational" crunching collapse, the original standing wave pattern is no longer reproduced. The box's width is not any rational fraction of its original width and the pattern is no longer as it was. As time goes on, the pattern becomes more jagged and only resembles the original pattern at certain repeatable periods. Here the self has gained the knowledge of the soul imperfectly. The message is garbled. But some of its essence is present. This corresponds to the soul's attempting to communicate with the self after it has already achieved a certain state of awareness. This could also correspond to life between birth and death and the difficulty each self has receiving soul-talk after birth.

enough, when such a collapse occurs, the pattern is never reproduced except for fleeting periodic occurrences. One might think of this as a distorted message from the soul, or, if the message is remembered, that is, put back together by the self, the message indicates there is something else beside the self and the world.

Irrational collapse garbles the message, although some of its essence is present. An irrational message still provides guidance, but since the message is not clear, it is often ignored. This explains the difficulty each self has receiving soul-talk. Such garbled messages could be heard normally by the mature self, but it would take some discipline to really listen to them. For example, soul-talk may be heard while meditating.

Finally, when the self wishes to communicate with the soul, the boundaries of the box undergo a sudden expansion. The pattern of the self, whether in the form of a garbled message originally from the soul or in a rational

Figure 13.5. Quantum Soul-Talk (D). After the rationally contracted box (B) has undergone a sudden expansion, the standing wave pattern is also not reproduced as it expands to fill the volume. Regardless of whether or not the box undergoes a rational increase over its original width, the pattern is never reproduced exactly. As time goes on, the pattern sloshes back and forth never resembling the original pattern. Here the self attempts to communicate with the soul by undergoing expansion. The self-knowledge is lost in the expansion. This could be a model for enlightenment or a model for what happens to the self or ego upon death.

vibrational pattern, suddenly expands and there is, associated with the expansion, a feeling of release or letting go. At such a moment the self feels released from tension. It's a bright new day, suddenly you feel good again. This could represent freeing the soul from physical bounds at the moment of death.

When the self talks to the soul, the box expands. In such a moment the self feels release from tension and self-less communion with God and all sentient life-forms in the universe. Buddhists call this nirvana. Christians call it faith or consciousness of God. I call it realized compassion.

Quantum Physics and the Two Time-Streams of the Soul

We have seen that the mechanism of soul-talk involves the collapse and expansion of the boundaries that limit the wave pattern of sacred information. Soul-talk involves a form of motion that goes beyond our normal conception of movement, namely, the setting of instantaneous boundaries in space, limits that bounce the wave function back and forth, forming standing wave patterns of information. Although this seems strange, it is no stranger than the collapse of the wave function in quantum physics—equally unexplainable by modern physics—that converts possibilities to actualities.

But what about time? Soul-talk seems only to involve space, so how does time enter the equation of the soul? Let me answer by asking another question. Since the soul acts as a guide to the self, where does it get its information? Here the work of Rudolph Steiner[18] and Qabala can help us to understand what the soul communicates to the self and how time enters the equation of the soul. One possibility comes from regarding the soul as living in the future or coming from the world yet to be.

Let me first put it briefly. The soul's voice travels on faster-than-light-filling-the-universe quantum waves of possibility both forward and backward through time. Normally, these oppositely-traveling-through-time wave patterns—God's Mind Dreaming—cancel out or reinforce each other as random dream patterns. These flashes of reinforcement do not amount to much—maybe the sudden and random appearance of particles that soon vanish into the emptiness of vacuum space. But when communication arises, the present self and the future soul emerge from God's Dreaming Mind. This is akin to God having a lucid dream. This reverberating *action* to *being* motion gently creates the sentient universe. In this way God becomes the universe, yet remains outside it. In a similar manner the soul becomes the body while remaining within the vacuum.

Figure 13.6. Faster-Than-Light Waves of Future-Seeking Possibility. Traveling faster than light and filling the universe, these are future-seeking possibility (FSP) waves as seen by a particular time-bound observer or self. They are shown as waves pointing to the right to indicate that they are all traveling forward through time.

To grasp the space-time dynamics of soul-talk, consider that the universe is filled with quantum waves of possibility traveling everywhere. You might imagine this to be the unconscious or, as I mentioned earlier, the dreaming mind of God. From one observer's perspective these waves, although traveling at speeds beyond that of light, are considered to be traveling forward through time, seeking their future. These are called future-seeking possibility (FSP) waves.

However, such waves are not limited by any time sense, since they are faster-than-light waves.[19] The direction of time is not specified by waves traveling superluminally. Hence, other faster-than-light-filling-the-universe quantum waves are also traveling backward through time as seen by this same observer. These are called past-seeking confirmation (PSC) waves.

You can think of these FSP waves as originating in the past, as far back as the big bang (the alpha point), marking the beginning of time, and the PSC waves originating in the far future (the omega point), the so-called end of time. The wave patterns of possibilities are essentially mirror images of each other and normally, like a positive and a negative, cancel each other out. But

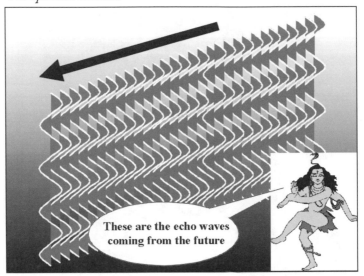

Figure 13.7. Faster-Than-Light Waves of Past-Seeking Confirmation. These waves of past-seeking confirmation (PSC), traveling faster than light, are time-reversed traveling backward through time as seen by the time-bound self. They are shown as waves pointing to the left to indicate their ability to interfere with and possibly cancel FSP waves.

there are exceptions. When an exception occurs, an event takes place—something happens in the universe. God's mind, dreaming of possibilities, creates a real event.

A great desire for communication then arises. An event in the pesent, the appearance of a self, and an event in the future stimulated by the soul, emerge from God's dreaming mind. Now the flowing waves develop an unexpected pattern. It must arise as a double flow of waves, one generating the other in a novel manner. One could picture the primary pattern running from the present to the future, corresponding to a message coming from the self and heading toward the soul and generating, in the future, a secondary pattern, matching the primary in intensity, but running backward through time to the source of the primary. This corresponds to the message coming from the soul and heading toward the self.

These time-reflective wave patterns must reinforce each other in a peculiar manner. They reinforce each other in the time interval between the present source and its future counter-source, and they cancel each other out for all time other than in this vital interval. This is seen as the movement from action (the self) to being (the soul) and back again to self and back again to

Figure 13.8. The Overlap of FSP and PSC Waves. When these time-reversed PSC waves overlap with the forward-through-time FSP waves, they mostly cancel out or weaken everywhere.

soul, continuing to bounce backward and forward impervious to time and at the same time gently creative of the temporal plane of existence.

Creative? Why? This pattern is not fixed and solid. It is not a sure thing. There is assuredly an element of chance and circumstance, a smidgen of the unknown, a creation of something new, something unexpected. It is a cosmic crap game, a throw of God's dice, a bet, a probability that what happens now will generate what happens next and that the two events, *now* and *next*, will be meaningfully connected.

Gentle? How so? We are not dealing with a solid thing. It is an action as soft as falling snow but as powerful as an avalanche, once the events *now* and *next* become strongly correlated and the probability edges ever closer to the ultimate gambler's dream: the number one—certainty.

The soul is capable of directing the action from the future, but it is vulnerable to its own loss of identity as it brings more of itself into the self-imagery process. But it must do this if the body-self is to evolve. Full self-realization comes when the two parts can talk freely to each other and see they are the same, just seen from different vantage points.

The soul ideally directs the body-self without ultimately becoming the body by remaining within the vacuum of space. But to communicate, it partly becomes the body and takes its form while it knows that it is not. It accepts its

Figure 13.9. The Creation of Karma: Cause and Effect. Here many events, indicated by the arrows, throughout time, alter the universe. The FSP and PSC wave patterns cancel everywhere else in the universe while they reinforce themselves between the events. The earlier events are taken as causes for the later events taken as effects.

own existence and it accepts that to move matter around it must partially become matter by literally jiggling the body's electrons around. It has an agenda, but this can only be experienced by the bodymind when the body's awareness of need and suffering are fulfilled.

Think of the soul as sending signals coming from the future. Think of the bodymind or self as receiving those messages. If the self is tuned to the soul's frequency, the self receives the message. Usually the self is preoccupied with nonconscious and mechanical survival and only gets the message once in a while. This preoccupation gives rise to the duality of all things as seen by the self. The soul attempts to guide the bodymind while the bodymind, in its preoccupation, attempts to fulfill its material addictions.

The game of the soul is the ongoing evolving creation of self and nonself without losing itself in the battle. It is to shape the self in the world but not lose its character of soul in doing so, even though the self embedded in matter demands it do so. This is the age-old problem, the war with time, as Carlo Suarès labeled it. The problem of the soul is intoxication. It can lose its purpose if it identifies too fully with the self.

However, the desire to become the body is very, very strong. The soul really loves the body and loves with all of its heart the self that is created. Remember, to the soul, the self is like a child. The soul can see what is ahead of the self, but it cannot fall into the self too closely or it loses its identity as soul. Think of the soul as a loving parent to the childlike self. The soul knows that matter is addicting because it is so drawn to the self—the soul's reflection in matter.

The Mothers of Creation

Although I've attempted to draw a picture of the soul in modern physics terms, I owe much of this insight to the ancient viewpoint put forward by the Chaldean and later by the modern Jewish mystics who painted the soul in Qabalistic colors. In chapter 4 we looked at some of the Hebrew symbols and their meanings. The Qabalists have long recognized that there were three symbols that were considered primary to existence and the nature of human consciousness. These symbols are called the "mother-letters." They are *aleph, mem,* and *sheen.*

According to ancient wisdom, the spirit or vacuum or God or aleph, first breathed a cloud into the universe. This cloud resisted anything else that God or aleph might produce. This resistance allowed whatever else God was to breathe to reflect back. When this reflection took place, consciousness of the universe occurred. When I met with the Biblical scholar Carlo Suarès, he explained that universe is spirit projecting itself, tending to become aware of itself, by emanating a cloud of consciousness upon which it can self-reflect. Gradually two energies appear—one coming from the source of consciousness and the other emanating from the cloud as a reflection of itself.

Suarès continued:

> What I have been trying to do is imagine how consciousness projects itself and what is the maximum possible energy [it can have] in any given universe. . . . I imagine first a consciousness that tends to become aware of itself and emanates, projects a cloud. That cloud may be mem, . . . the least possible energy. That is what your quantum may be, mem, . . . the projection of the most inner state of contradiction of consciousness that is going to create something that will not be itself, but will allow itself to realize itself.

I asked Suarès how it all worked mechanically. How could consciousness project a cloud? Doesn't that imply awareness already to do this? Accordingly the other mother letter, *sheen,* which is also pronounced *seen,* enters the arena. Qabalists see sheen as the breath of God or as the spirit itself. It resembles the action of the two—PSC and FSP—streams of quantum waves of possibility described above. Suarès answered:

> I don't know how it happens, but it does happen. It is not aware; it senses an inner necessity. A cloud of consciousness appears which is not aware. Gradually two energies appear. One is the energy that comes from consciousness. The other emanates from the cloud, the projection, as a reflection of itself. One is sheen and the other is seen. Thus, I write the equation:
>
> aleph = mem × seen × sheen.

He made a drawing of this process. It symbolizes for me the basis of human consciousness in one simple but strongly significant picture.

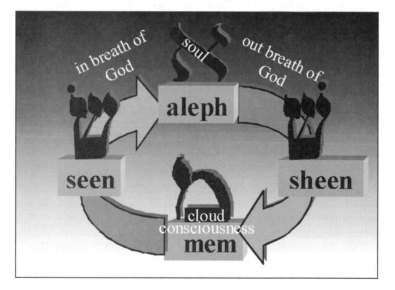

Figure 13.10. The Mothers of Creation. Aleph breathes into mem causing the universe to come into being and consciousness to arise simultaneously. The breath of God returns to aleph.

The three "mothers" of creation, aleph, mem, and sheen, work this way. The "breath of aleph" blows as sheen into the "waters of consciousness," mem, and structures are created. These appear as seen in the constricted form of this breath which returns to aleph.

These two flows or processes occur simultaneously, seen coming from mem and sheen coming from aleph. Suarès continued:

Here I find that both seen and sheen are conditioned by existence. What existence? The existence of the projection of the psyche and the existence of the psyche itself, which comes into existence through this process of the two energies.

I remembered that sheen and seen are both spelled *sheen-yod-nun*.[20] Thus when Suarès says that they are conditioned by existence, he means that their movement is through the letter yod—the symbol of existence. The goal of that movement is *nun*-final,[21] the principle of cosmic indetermination. Thus, the actions of these two invisible movements do not create any final or set condition but free and set new conditions through indeterminism. Existence, the letter yod, is at stake in the middle of all this. Since nun is also the symbol of life in existence we can also say that sheen acts through life as it actually exists in plants, animals, and humans. We evolve through the recognition of the cosmic indetermination of life. *Life's only goal is not to repeat itself endlessly.*

Prayers To and From the Future

According to Rabbi Micha Berger of the Aish Das Society, the sacred prayer known as Tephilla indicates God's use of time-reversal. God refers to both *causality* and *teleology* constructing the universe from every past event as *cause* (a reason) and every future event as *telos* (a purpose). Rabbi Berger states:

> What most readily jumps to the eye about the Tephilla is that the twenty-two words it opens with are an anagram of the Hebrew alphabet in reverse. While many [prayers] are written with an alphabetic motif, it is far more rare for the alphabet to be presented in the reverse. What concept were the authors trying to express with this sequence?

According to Scripture God says, "I am the first and I am the last; and besides me there is no God, who is like Me," a sentiment found a number of times in the Tephilla. What could God mean? "First" and "last" can refer to a sequence or to a temporal order. But what sequence does God refer to?

The rabbi believes that God refers to causality and teleology. In other words, the universe is constructed from the fact that every event has a cause, an event that preceded it and forced it to happen, and a telos, an event that followed it and is the purpose for it. He goes on to say that although teleology is in disfavor today (coming out of the era of Darwin, when life was seen to be the product of accident), for the spiritually minded there is no question that

God created the universe for a purpose, and the purpose will be met. Rabbi Berger goes on:

> Thus everything has two reasons for happening: Its cause and its purpose. When we realize that everything that happens to us is for a purpose, everything is part of that pursuit of the Culminating Purpose, then we are at peace.

Like the Tephilla, one can think of prayer as reaching out to the future. Soul-talk responds when the prayer is reversed so that the time-reversed message and the prayer coincide.

Do You Have a Soul?

Now that you have a complete picture of how soul-to-self communication occurs, I'll answer the burning question: Do you have a soul?. Of course you do: the same soul that I have. How do you know you have one? Because you are conscious, experiencing the one same mind that each of us experiences. That awareness is your soul reflecting to your self, sending messages by continually collapsing itself to fit your boundaries. You can communicate with it at any time by being conscious of the world around you, and expanding your boundaries. This is called surrendering your ego, letting go, getting off your high horse, your royal "I-ness." By sensing the world, you communicate with your world-soul, the world, and the universe.

You can also lose your soul at any time by going on automatic addictive pilot and behaving nonconsciously toward everything including yourself. What happens to a world that behaves this way? What happens to a person who behaves nonconsciously? Since there is no awareness of soul, no responsibility for anything can occur. The world becomes lost in an illusion of uncaring mechanical action and reaction. Since this is an illusion, suffering mounts to unbelievable heights and violence erupts. A soulless person responds in the same way. A soulless world is nearly unthinkable, but the danger of the world becoming this way is always present. Soul awareness must be brought into the intellectual world of science just as intellectual responsibility must be brought into the world of religion.

In the final chapter we'll look at some right answers to what I consider to be responsible questions. I'll offer definitions of the soul, self, matter, and spirit based on my new vision of science and spirituality. We'll also take a look at some answers raised when science and soul are brought together and what takes place when your soul and your self have a conversation.

CHAPTER 14

Some Soulful and Right Answers

Seeing into Nothingness—this is the true seeing, the eternal seeing.

—Shen-Hui, 8th century

In appearance I'm a thing moving about in Space. In reality I'm that unmoving Space Itself.[1]

—Douglas Harding

He who knows that he is Spirit, becomes Spirit, becomes everything; neither gods nor men can prevent him . . . The gods dislike people who get this knowledge . . . The gods love the obscure and hate the obvious.

—Brihadaranyaka Upanishad, 7th century B.C.E.

THIS WAS NOT AN EASY book for a scientist to write. An uneasy feeling still resides within me after my attempt to model the soul using scientific concepts. Perhaps my own trickster-consciousness plays tricks with me. Perhaps there is a deeper reason for my uneasiness. Since the soul is a process and contains no mass or energy, what could a scientist have to say about it? The action of the soul is not predictable. It cannot be controlled. It is imaginal. Science attempts to deal with what is real, material, and predictable. Yet it appears science may be fooling itself in its attempt to render all phenomena as rational, predictable, and objective. Science may be dealing with an incomplete deck.

I am certainly not the only physicist who believes this. Physicist Wolfgang Pauli in writing about Kepler once said, "Scientists went a little too far in the 17th century."[2] This was Pauli's way of saying that we have adopted a far too rational approach in our attempts to understand nature. Pauli recognized the unconscious was far more instrumental in making theories about matter than most physicists would have even contemplated. Since science was created to enable human beings to cope with and ultimately control nature, the shadowy trickster side of that unconscious surely popped out. In its attempt to rationalize experience, much of modern science has taken humanity on a logical but heartless venture. Thus Pauli was convinced that a new conception of reality had to include spirit and matter as complementary aspects of one totality.

Where does science's error occur? Every equation we write down as scientists, every object we see and classify, put into logical order and rearrange when that order changes, ultimately boils down to our fleeting sense impressions. Mistakenly, then, science deals with classification of inert nouns, while reality appears to be entirely made of active verbs. This activity, the recognition of unpredictable process, involves consciousness every step of the way, and this activity eludes mathematical description and so far eludes science.[3]

Spirituality deals with this activity as something subtle, vital, immaterial, unpredictable, and conscious. In Christianity, Judaism, Hinduism, and Islam this activity is called the essential soul. In Buddhism it remains unnamed. Certainly Buddhists deal with consciousness as arising in interdependency with the objects of consciousness—the images presented to the sixth sense, the mind. However, the essential "whatness" of being aware remains a nameless activity, "a phantom, dew, a bubble, a dream, a flash of lightning, and a cloud."

Present science, based on Aristotelian and Newtonian models which lead to rationalistic, reductionistic, materialistic, and nounlike thinking, at best believes the soul to be no more than an epiphenomenon. The intangibility of the soul restricts science at worst to regard it as imaginary—an illusion of our thoughts and senses. Either way it would incorrectly reduce the soul and consciousness to purely physical and mechanical energy. I tried to show why that viewpoint, based as it is on the old physics, is wrong. With quantum physics, we see why this error in thinking occurs. The effect of consciousness on physical reality is not accounted for in classical physics.

Throughout this book, we have dealt with reality and illusion. These issues arise from considering what questions to ask about the soul. In our earlier attempts to define the soul, I suggested we were asking the wrong questions and hinted there were right questions to ask. What is the essence of a wrong question with regard to the soul? What is a right question in this regard?

The answer lies in the concept of *objectivity*. In attempting to define the soul objectively, we are forced to engage all the mechanisms that distinguish one object from another. These mechanisms seek out differences between things. But in dealing with the soul, objective evidence is not and will not become available. The soul cannot be separated into parts nor into individual souls; that, too, is an illusion brought about by attempting to view the soul objectively, to count it as one counts bodies. The soul is real and subjective; no one can demonstrate its existence as objective fact. Questions attempting to objectify the soul are simply the wrong questions. Questions dealing with the spontaneous activity of the soul are the right questions in spite of the difficulty in finding answers and the lack of substantiating evidence for those answers. Can we find the right answers to these questions?

I believe the answer is yes, however the scientist in me still feels uncomfortable. Scientists realize that truth is extremely elusive. What was scientifically true one hundred years ago is mere error today. Anything I say about the objective qualities of the soul could suffer the same fate. Yet, as a scientist I am compelled to make the attempt. When a better vision or model comes along, I shall be quite happy to see mine replaced. Since there hasn't been a model of the soul since the seventeenth-century days of Kepler, I shouldn't be too remiss in placing one before your eyes.

In all of my soul-model making I need to look into the heart of my inquiry. I need to engage spiritual metaphor, feelings, and fleeting moments of inspiration to do this. This means learning to use a different way of seeing into the affairs of the material world. This requires separating reality from illusion.

What is real and what is illusion? I conclude, in agreement with the Buddha, that the self and any of its material interests, in spite of how compelling they are, are illusions. The soul is real in agreement with Buddhist thought and its deeper implications as I presented them in part 3. Our present worldview reverses these and follows the diagram of Plato's more inverted line (figure 3.5) as its banner. This view values opinions over knowledge and puts far too much importance on what is visible rather than on what is intelligible. The domain of illusions plays a role far greater than the domain of reality. It is time to reevaluate and reverse the priorities exhibited on Plato's more inverted line.

Science being the creation of mind following this inversion has been led astray and mainly used to satisfy material wants. Hence, science through its partner, industrialization, has led us into a subtle trap of our own making—a trap that has led to a world in which less than 10 percent of its population survives well while the remaining 90 percent lives in near poverty.[4] This trap confines the soul to the body and puts forward the idea that nothing survives beyond the body. Survival becomes the major issue and exploitation becomes the servant of science. The logical consequence of such a worldview is the satisfaction of material wants by the few at the expense of spiritual needs, the lack of material by the many, and the creation of a needy, materially addicted world.

To see through this illusion, the soul must be included in *everything* we do. Every transaction must be compassionately soulful in awareness of the suffering in each sentient life-form. I don't mean anything gloomy by this. Simply remember the soul is present in every human being, in every animal, and in all life-forms including our own planet. To me, this is the right answer to every question: Remember, your soul is everywhere.

Regardless of its name, the soul is a natural process involving *consciousness of knowledge*. This process occurs in the vacuum of space, began when the universe first appeared, and will end when the universe returns to the void. Thus the soul follows the same path as matter/energy, appearing during the big bang (birth) and eventually disappearing at the time of the big crunch (death). Since the soul begins when time begins and ends when it ends, the soul is immortal: It lasts for all time. However, unlike matter, it is not objective and cannot be controlled as a physical object. It has no mass or energy. Yet it is real.

Although I can only deal with the soul through my limited thoughts and experiences, including those of a physicist, I believe my new vision of the essential quality of soul and its relationship with matter, spirit, and the self are

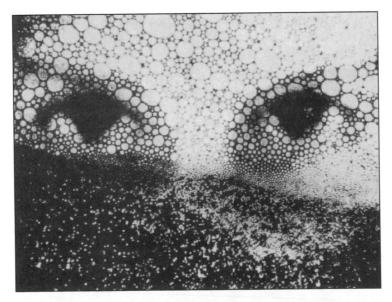

Figure 14.1. The Soul is Everywhere You Want To Be.

useful. However, I am not fully comfortable with such definitions. I realize how I, too, as a scientist may be fooling myself as a I suggested science has fooled itself. Nevertheless, a new model that brings the soul into science certainly is needed. Perhaps from it a better bridge connecting science and spirituality can be constructed.

My concept of a "new physics of the soul" shows how the soul, the self, matter, and consciousness are, although related, not equivalent. The soul manifests matter and energy and shapes the material world by knowledge. In what follows I will provide some definitions explaining how the soul enters the temporal world and through knowledge shapes matter.

Definitions of Spirit, Soul, Matter, and Self

We have taken a far-reaching tour dealing with soul, spirit, self, ego, matter, energy, and so forth. Next, I provide some important distinctions and definitions of these concepts. When reading the following definitions I want you to picture a guitar. Imagine one of its strings. The top of the string is bound at the neck. The bottom of the string is bound at the base. The string can be held against the guitar's fretted fingerboard. When the string is plucked it vibrates. By changing the placement of the fingers on the board, thus holding the string against different frets, different notes are heard from the plucked

Figure 14.2. The Soul Is That Old Sweet Song We Know So Well.

string. The places where the string is held down are called nodes. No matter which note is struck, the nodal points remain fixed. The string thus only vibrates between two nodes: one at the base and the other at the point of contact of the finger with the fret or, when unfretted, the neck.

Spirit

In this book I have used the term *spirit* to mean "the vibrations of nothing." I know of no way to accurately imagine such vibrations. The closest I can come is through the notion of aleph (the potential to become anything). The vacuum is alive with these vibrations. They contain the potential for anything. The process for realizing anything, becoming aware of anything, results from reflection of these vibrations in the form of waves. To accomplish this reflection some form of resistance that bounces these waves back in the direction from which they came must arise. How such resistance arises is unknown. Carlo Suarès imagined this resistance as a cloud. Consequently spirit remains simultaneously potentially able to be aware and potentially able to be unaware when unreflected (that is what is meant by the *unknown* spirit, it means the *unreflected* spirit). The unreflected spirit is like an infinitely long vibrating guitar string with no nodes that shimmers in the wind.

Soul

When reflected temporally (not necessarily temporarily) at *nodes of time*, spirit becomes partially aware and partially unaware. We call such a reflection *soul*. There are two kinds of nodes: *temporal* and *spatial*. A *temporal node* is a point in time and a *spatial node* is a point in space. Such points are the markers of events. Without such markers, the very notions of space and time are meaningless. Thus, I am posing that temporal markers, even though they imply finite duration, are necessary for creating the immortal soul out of spirit. Similarly, spatial markers defining extent, length, area, volume and such, are necessary for the creation of matter, again from spirit. I'll get to matter shortly.

Thus, through temporal reflection, soul-awareness takes place along with its creation. Although the soul is eternal, paradoxically it arises from reflection of the spirit's vibration at two temporal points (like the nodes of a guitar string). This may seem contradictory since this would time-bound the soul, which is supposedly immortal. The resolution depends on at which temporal nodal points the reflections occur (like the guitar string vibrating with different fingering of the frets). These paradoxical reflections occur at the omega point of our history, the so-called end of time or big crunch (the end of the neck of the guitar), and travel back through time to our present period. They occur at the alpha point, the so-called start of time or big bang (the base where the other end of the string is fixed), and travel forward through time to our present. Thus there is no paradox, the soul is immortal lasting through all time. For before alpha and after omega there was no time, for time had no meaning (the guitar string was infinite and shimmering in the wind).

In this way an Einsteinian Relativity of the soul manifests. Thus, although from our vantage point soul appears temporal, from its own viewpoint it appears immortal; it lasts as long as time does. The reflection of spiritual vibration beginning and ending at alpha and omega—the nodes of time—makes the soul conscious (the unfretted but bound guitar string vibrating in its fundamental tone or any of its harmonics). Remember, although soul is conscious (the string vibrating in an orderly fashion as if it were plucked), it remains potentially unconscious (the string vibrating in a disorderly fashion shimmering in the wind and bombarded by random fluctuations of temperature) as does spirit. This means the soul can under some circumstances be rendered unconscious. When the soul is polluted (the string grows thicker and harder to pluck), which occurs when the body because of its own material pollution is no longer attuned to it, it falls into unconsciousness (the thickened string vibrates randomly more than harmoniously).

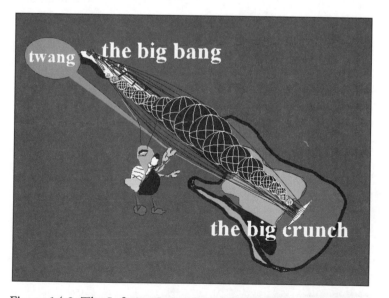

Figure 14.3. The Infinite Guitar String of the Soul. It vibrates between the nodes of time: the big bang and the big crunch.

Matter

Spirit can also reflect from the beginning and ending of space. When reflected at *nodes of space* (I'll also explain what these are shortly), spirit is unaware but potentially aware. We call such a reflection matter. For alpha and omega are truly the nodal points of both space and time. *The reflection of spirit in space is matter.* Spatial reflection of spirit or matter is unconscious or unaware. (Think of a still photograph of the guitar string vibrating. The vibration can be seen, but not heard!) Therefore, the spatial reflection of spirit from alpha and omega creates the material universe (photos of vibrating strings). However, while matter is unconscious, it, like the spirit, contains the potentiality for awareness. (Think of a movie of the vibrating guitar string. Now sound can be heard since the motion can take place in time.) That means under some circumstances matter can become aware! We see matter aware in every plant and animal and even (sometimes!) in ourselves.

The different forms of matter arise from the varying reflections in space of unaware patterns of spirit (photos of different vibrational patterns of fretted strings). All particles arise as variable reflective distances or in the language of quantum physics, variable wavelengths. When the universe was in its infancy, and very tiny, it trapped some of the vacuum vibrations inside of it. A trapped vibration became a particle. As the universe expanded, it outgrew the tiny trapped vibrations and they persisted as particles of matter. Thus, these different particles emerged according to their individual wavelengths. All particles are self-reflective unaware patterns of spirit. Such forms possess inertia. Inertia is evidence of unconsciousness or unawareness. Inertia is resistance to time passing.

Unconscious behavior is mapped as physical laws: The laws of science are the prime example. These laws describe which reflections are allowed and which are forbidden. Allowance and forbiddance are determined by balancing waves with respect to spatial boundaries. In the subatomic realm these laws constitute the field of particle physics.[5]

Since both the beginning and ending of time take place at a single point, we have the paradoxical notion that matter arises from reflections from points that have no spatial separation even though they are temporally separated by 20 billion, billion years of time. This is nearly unthinkable. As the big bang unfolds, space itself is created starting from a point and spreading out to the "edge of space." As the universe continues expanding certain patterns of reflection become stabilized. Some particles seem to have finite size or wavelength. Some, such as the electron, reflective of the starting and ending points, seem to have no size at all.

Self

The reflection of soul in matter is self. (The self is like the movie of the vibrating guitar string. The soul is like the string itself. After running the movie through a projector from beginning to end, the film is worn out and can't be replayed. Meanwhile the guitar gently sings.) Thus, self is a reflection of a conscious spirit (soul) in an unconscious but existing-in-space matrix (matter). Self is a conscious reflection of the soul from an unconscious reflection of spirit. (Think of the self, unmindful of its soul, watching the movie of the singing guitar.) Since it consists of conscious and unconscious modes of reflection (the string and its movie-image vibrate together), self is in part conscious and in part unconscious. Self unconsciously aligns with the bodymind, that part of the physical world that minds or governs the body. The origin of

Figure 14.4. Definitions. As long as the reflective process is purely temporal and nonspatial, the soul remains in the vacuum, free and aware, but highly susceptible to becoming physical, that is, falling into intervals of space and time. When the reflection of spirit's vibration becomes spatial, meaning the formation of matter, the soul becomes enamored with its reflection in matter and tends to become self. We say that the soul has been smitten. It then falls into space, time, and matter. The soul then experiences birth and death—events that are illusions for itself. (Birth and death are the start and end of the movie, not the guitar string which vibrates forever)

the unconscious in all human beings and the origin of the interplay between conscious and unconscious behavior in people follow from the mix of conscious and unconscious spirit in that reflection in matter (listening to the movie guitar and ignoring the real one).

Soul and Self: A Conversation

While researching this book, I was struck by the fact that ancient writers seemed to take the subjects of their writings in a far more lyrical or theatrical fashion than we do presently, often inventing fictional characters to flesh out abstract concepts. Galileo used dialogue to present his views on the "new science" of mechanics, and Plato is well known for his dialogues regarding morals, ethics, how to run a city-state, and even how to face one's own death. In such works the dividing line between fact and fiction is crossed many times.

The relationship of self and soul is often a heavy subject, so I will borrow from my ancient forebears, especially Plato, and present some *high-level* dialogue dealing with some important issues concerning the existence of the soul. In reading this it might help to think of the soul as reflective consciousness potentially able to be unconscious (a plucked vibrating string becoming an unplucked shimmering string) and the self as habitual addictive unconsciousness potentially able to become aware (a movie of the string capable of interacting with the real string).[6]

Keeping the above in mind, think of the guitar and imagine that your self and your Soul are having a conversation about their relationship.

> SELF: *So what is my relationship with you? How do I become aware of any relationship with you? In other words, how does my self become conscious of my Soul?*
>
> SOUL: *You have a very intimate and imaginal relationship with me. I come to you at various times, particularly when you are considering earth-shattering questions about the meaning of life. You are not equipped to handle the really big ones, you know. In fact, most of the time you live on what you might call automatic pilot. But things happen that you can't account for. Then, all of sudden, they happen again and something within you gains a definite opinion about it. You may not like it. Or you may become quite attracted to it. In fact, there are times when you become positively addicted to the things that happen to you.*
>
> *For example, you are very interested in yourself. You like to look at yourself in the mirror. You love to hear your own voice. You enjoy the*

smell of your body even when it doesn't smell too good to others. You really delight in you. In fact, if a day goes by when you haven't been able to enjoy your self, you get frustrated and a little despondent. If several days go by, it gets much worse. Finally when you get a chance to luxuriate in your self again, you feel better, particularly if you think that your enjoyment feeds your Soul as well as your body. You guess it's I, your Soul, that you feed, but perhaps it's really just your self. You're not sure at all.

You see there are days when you hate your self. You don't know who or what to turn to. Nothing you do for your self seems to make any difference. You feel lousy. You then feel your Soul has left you, and you doubt if you ever had one.

You see you don't believe that you have an involvement with the Soul as if it were a real person. If you were involved with the Soul, you wouldn't constantly be swinging from elation to despair, either stuffing your face with everything nice—and scouring the planet to do it—or denying your humanity by putting your body on a starvation diet. Or, even worse, if you felt your Soul's presence, you wouldn't attempt to kill your humanity by asking one part of your self to destroy another. It's your Soul that saves you from all of this self-reflective destruction.

SELF: *I don't quite get it. Do I have a Soul, or not?*

SOUL: *It is a bit difficult for any of us to get it. See, your Soul is a reflection of you, or rather, you are a reflection of me. I am not there right now, but you are. Or, in a way, I am there but I speak through you, if you let me. Often I'll speak to you in your dreams, sometimes when you are meditating. Sometimes you hear me when you are traveling. You always know when I am present because you feel worthy inside, really good, and you've done nothing physical to your body to make yourself feel that good.*

I, your Soul, always have your best interest at heart and know about everything you do. I don't make you suffer, but you make your self suffer at times, and you don't know why you do that. When your Soul speaks to you, you don't hear words, you have feelings. When you feel your Soul's presence, you generally feel as if a great weight has been taken off your chest. I have lifted it. You feel as if you have expanded your awareness. I have made you aware. You know after a Soul-to-self talk, you are more than you think you are and you also know whatever you think you are is always an illusion.

SELF: *Let me see if I understand you. I feel my Soul, but I don't see it, speak to it, or hear it. Is this correct?*

SOUL: *You see me whenever you see a newborn child. You talk to your Soul whenever you speak to a five-year-old. You hear your Soul speak when you listen to an old woman tell you about her life. You hear your Soul sing to you in a Beethoven symphony, in a Philip Glass composition, or a song by Ray Charles. You read your Soul's words in the lines of Hamlet, "To be or not to be," or in the words of Eldridge Cleaver in his book* Soul On Ice. *You hear them in a children's poem. You feel the Soul whenever you sense the beauty of the world or its sadness.*

SELF: *My self certainly understands what you are saying, but this still doesn't prove that you, my Soul, are anything more than a romantic illusion. Could you explain how I can know you exist?*

SOUL: *You are not adequate for the job. You don't know that your Soul exists as long as you insist you do! All you know is what you feel when you feel the presence of something sacred within you that can't be explained by mere satisfaction of physical sensation. You, like all of humanity, feel your Soul's presence when you feel compassionately aware of each other. That is enough for you.*

SELF: *So, if I hear you correctly, the Soul is more than comforting thought. I have never seen my Soul. In fact, I have never even heard the Soul speak to me as you are now.*

SOUL: *The Soul does not appear to you as an object in space. Nor does it speak to you with a voice separate from your own. In spite of all that, Soul exists. To call the Soul an illusion may be correct from a certain materialistic position. But according to your understanding of such things as quantum physics, the material world stands on shaky ground and so cannot be trusted to provide the foundation of reality anyway. It certainly should not be used as the basis for the evidence of my presence.*

SELF: *Do you have any advice you can give me to help me cope with the world, life, death, and all of my problems?*

SOUL: *I do not have a material form. I am imaginal and therefore the basis for all of reality. Humanity needs to listen, until such a time as the voice of the Soul is heard throughout the universe as the only voice of compassion and reason that has ever existed. When this occurs, all of humanity will be truly free and the voice of the Soul will sing until the end of time.*

Notes

CHAPTER 1
Some Soulful but Wrong Questions

1.   We examine one of these scientists' viewpoints in chapter 5. See Frank J. Tipler, *The Physics of Immortality* (New York: Doubleday, 1994), pp. 1, 2.

2.   See Steven Weinberg, *The First Three Minutes* (New York: Basic Books, 1977). He ends his epilogue with his feeling that "the more the universe seems comprehensible, the more it also seems pointless."

3.   See Fred Alan Wolf (12), "The Quantum Mechanics of Dreams and the Emergence of Self-Awareness," in *Toward a Scientific Basis for Consciousness*, eds. S. R. Hameroff, A. W. Kaszniak, and A. C. Scott (Boston: MIT Press, 1996) and David J. Chalmers, "The Puzzle of Conscious Experience," *Scientific American* (December 1995).

4.   Michael Grosso (2), "The Status of Survival Research: Evidence, Problems, Paradigms," *Noetic Sciences Review*, no. 32, (1994), p. 12.

5.   Francis J. Crick, *The Astonishing Hypothesis: The Scientific Search for the Soul* (New York: Charles Scribner's Sons, 1994).

6.   The *quantum wave function* is a mathematical formula that presents the possibilities of events occurring in the form of a wave pattern distributed through space and changing in time much as an ocean or sound wave ripples and flows. However, as seemingly physical as this appears, it has no physicality in and of itself. Moreover, it cannot even be represented by real numbers but must be represented by complex numbers consisting of real and imaginary numbers.

7.   One of the best discussions of the importance of virtual processes is found in Richard Feynman (2), *QED: The Strange Theory of Light and Matter* (Princeton: Princeton University Press, 1985).

8.   In 1961 while I was employed by the Lawrence Radiation Laboratory in Livermore, California, I published a paper with my colleague Dr. Marvin Mittleman on this subject. In doing the calculation I was impressed with this process of infinite goings-on by atomic electrons in the simplest processes. See M. H. Mittleman and Fred A. Wolf, "Coherent Scattering of Photons by Atomic Hydrogen," *Physical Review*, vol. 128, (1962), p. 2686.

CHAPTER 2
Aristotelian Soul Physics

1.   Of course, we can also ask about the seat of consciousness. No one has found it in the brain or anywhere else, for that matter. Yet we know that we are conscious. Soul-evidence must be seen in a similar light. We may not find the soul in the

body, yet like consciousness and the sense of presence of self we call self-awareness, we intuitively *know* it is there. I will say more about senses and knowing later.

2.   In this regard, look ahead at Plato's argument in chapter 3.

3.   Faster-than-light (FTL) travel is possible for a special kind of material called *tachyonic matter*—a theoretical substance without real mass but possessing *imaginary* mass. Peculiar to such matter is the ability to travel at infinite speed with no energy at all and to possess infinite energy at the lowest speed possible—the speed of light. This "confinement" of tachyonic matter to superluminal speeds produces weird time-travel possibilities wherein such matter is capable of moving backward through time with the whole "timescape" stretched before it.

4.   Even light can be made to go faster than light. This paradoxical situation arises when light tunnels through a barrier of slight thickness in the millimeter range. Apparently photons (particles of light) raced ahead of other photons when they penetrated the barrier. See "Faster Than What?" *Newsweek* (June 19, 1995).

5.   Photons are very strange indeed. *First*, they exist only as waves moving through whatever space they encounter when they are not seen! The minute they are viewed, by knocking into the retina of an eye, or a photocell, they are annihilated. *Second*, at that instant, the death of the photon, so to speak, the photon behaves as if it were a particle. *Third*, moving at the speed of light it travels through space and time, but for itself, according to relativity theory, it goes nowhere and no time passes at all. Consider this: A photon is always born or created in a subatomic process and is destroyed in another subatomic process. To observers of the processes, the photon travels from one process site to the other at the speed of light, taking whatever time is required. But to itself it goes nowhere and no time passes at all. Hence for a photon birth and death are simultaneous.

6.   Barbara G. Walker, *The Woman's Encyclopedia of Myths and Secrets* (New York: Harper and Row, 1983).

7.   We will examine this question in more detail in chapter 3.

8.   In chapter 11, dealing with the Buddhist concept of skandhas, we will look at thinking as if it were a sixth sense, like hearing or seeing.

9.   By "possess" motion he meant capable of motion or having the potential to move.

10.   This is an arguable point. A field, not being a body, can indeed cause a body to move. However, a field can be viewed as made of potential particles of matter or potential bodies. Hence when a field causes a body to move one could argue that the field's "bodies" are doing so. This is essentially the position taken by physicist Richard Feynman in his theory of quantum electrodynamics.

11.   So does the Buddha who takes it that the soul doesn't exist. Aristotle (384–322 B.C.E.) and the Buddha (550–480 B.C.E.) were less than 200 years apart. Perhaps this thinking was in the air.

12.   In some way, Aristotle, with regard to the soul, may have been anticipating the notion of a hologram, wherein each part contains information about all of the sources that were used to construct it.

13.   Think about this for a moment! The soul is immovable, perhaps like a great mountain. The body is movable only under the action of the mountain-like soul.

14.   That is, if it did not possess sight.

15.   Figure 2.1 may be helpful at this point. By initial actuality he means the left side of level one.

16.   Again refer to figure 2.1.

17.   Look at chapter 3, where we discuss these qualities as ideals rather than *actuals*.

18.   See my book *Taking the Quantum Leap: The New Physics for Nonscientists*, rev. ed. (New York: HarperCollins, 1989), p. 9.

CHAPTER 3
Platonic Soul Physics

1.   This quote comes from Bertrand Russell, *History of Western Philosophy and its Connection with Political and Social Circumstances from the Earliest Times to the Present Day* (London: George Allen and Unwin, 1961), p. 140.

2.   Morris Kline, *Mathematics: The Loss of Certainty* (New York: Oxford University Press, 1980), p. 9.

3.   I mean by this statement that in ancient Greece women were not held in very high regard.

4.   An electron had only a certain energy which was according to its orbital quantum number proportional to one divided by the square of that number. Thus an electron was an inverse square number manifesting as matter.

5.   For those interested in the geometrical properties of ancient Greek numbers, consider the sum of two consecutive triangular numbers always equaled the square of the larger number. Thus, for example, 2 triangled plus 3 triangled equals 3 squared, 3 triangled plus 4 triangled equals 4 squared, etc. While this is somewhat of a mystery to modern ways of thinking, it is easy to see why it is true when you look at the geometry of triangular and square numbers. Examine figures 3.2 and 3.3, and if you look closely you may see how it works.

6.   Marie-Louise von Franz, *Number and Time: Reflections Leading to a Unification of Depth Psychology and Physics* (Evanston, IL: Northwestern University Press, 1974).

7.   I have copied this drawing from Plato (2), *The Republic*, rev. ed. Translated by Desmond Lee (New York: Penguin Classics, 1987), p. 310. I have modified it to some degree, and the discussion of the meaning of the line is my own. The modifications of the line I use below in the next two drawings are also my own.

8.   Although I have used the ratio 3/5 or its inverse 5/3 purely by aesthetic choice, I want to point out that 5/3 is very close to *phi*, the so-called Golden Mean. Phi = 1.618 . . . , and is an irrational number often used by architects to please the eye in the construction of buildings. In a poll taken by Gustav Fechner, rectangles showing the proportion of the length of one side to the adjacent side with ratio phi and others were viewed by a large number of people. The rectangle using the ratio phi was greatly preferred over other proportions.

9.   Could this indicate evolution is not possible without death and reincarnation? I'll examine this in greater detail in chapter 11.

10.   This law is demonstrated in every chemical reaction. For example, when water is turned into gaseous hydrogen and oxygen, the reaction is signified by $2H_2O \rightarrow 2H_2 + O_2$. Due to the law of mass conservation those extra "2"s are necessary to tell us two molecules of water must be used to make two gaseous molecules of hydrogen

and one molecule of oxygen. Adding up the number of atoms of hydrogen on both sides of the reaction "arrow" we find exactly the same number, four. Similarly, there are two atoms of oxygen on both sides of the arrow. The formula predicts twice as much hydrogen gas is generated in the process (two molecules of hydrogen for every one molecule of oxygen), as anyone who has done the experiment in chemistry lab knows.

11.   One might be tempted to regard the nonphysical substance of the soul as energy. However, this is a mistake, as I shall show later. For Plato, the notion of nonphysical substance meant ideal substance, a substance having the perfected attributes of real substances. As we will see later, Plato might have been foreseeing the notion of quantum waves of probability—nonphysical, downright imaginal, but, nevertheless, ideal forms of possible matter.

12.   The principle of complementarity says attributes of the physical universe can never be known independently of the observer's choices of what to observe. These choices fall into two distinct or complementary pairs of observations called observables. Observation of one of the pair-observables always disturbs the other so that simultaneous observation of both is impossible. The tree and the forest are complementary pair-observables. If you observe the tree you cannot see the forest, and if you observe the forest, any individual tree cannot be observed. Similarly you can listen to a musical theme and not hear a note or you can listen to a single note and not hear the theme of the music.

13.   G. Sperling, "Information Available in Brief Visual Presentations," *Psychological Monographs* 74, no. 11, (1960). I described this experiment in Wolf (10), *The Dreaming Universe: A Mind-Expanding Journey into the Realm Where Psyche and Physics Meet* (New York: Touchstone, 1995), p. 308.

14.   I've taken much of this from Plato, "The Immortality of the Soul and the Rewards of Goodness," Part 11 of *The Republic*, rev. ed., trans. Desmond Lee (New York: Penguin Classics, 1987), p. 440.

15.   Plato would argue the bad food caused the body to manufacture the substances detrimental to its own nature and the food itself did not kill the body.

16.   In *Timaeus*, Plato argues the soul does contain lower parts that are mortal. Later I shall regard these mortal parts as the self generated by the brain/mind.

17. I'll take up the matter of good and evil once again in chapters 9 and 11, and specifically compare what Plato says about the soul not suffering extinction even though it suffers, with the ideas about hell expressed by Meister Eckhart in chapter 11.

CHAPTER 4
The Ancient Basis of a Modern Soul

1.   Francis Crick, *The Astonishing Hypothesis: The Scientific Search for the Soul* (New York: Charles Scribner's Sons, 1994).

2.   Frank J. Tipler, *The Physics of Immortality: Modern Cosmology, God and the Resurrection of the Dead* (New York: Doubleday, 1994).

3.   For other examples see Dannion Brinkley (with Paul Perry), *Saved by the Light: The True Story of a Man Who Died Twice and the Profound Revelations He Received* (New York: Villard Books, 1994) and Fritjof Capra and David Steindl-Rast,

Belonging to the Universe: Explorations on the Frontiers of Science and Spirituality (San Francisco: HarperSanFrancisco, 1991).

4.   Of course, I know the environment is being looked after by environmentalists, ecologists, and enlightened people everywhere. However, they are the exception today and not the rule.

5.   Such as those of Tipler and the Dalai Lama. But perhaps you find these views not as strange as I do.

6.   Bernard Grun, *The Timetables of History: A Horizontal Linkage of People and Events* (New York: Touchstone, Simon & Schuster, 1982).

7.   Karen Armstrong, *A History of God: The 4,000-Year Quest of Judaism, Christianity, and Islam* (New York: Knopf, 1994), p.11.

8.   This phrase comes from the Book of the Dead and states the first solitary appearance of Atum. Note the change of tense in the sentence. It implies that Atum always exists and therefore only can be "I am." This theme will appear again in this chapter. See Henri Frankfort, Mrs. H. A. Frankfort, John A. Wilson, and Thorkild Jacobsen, *Before Philosophy: The Intellectual Adventure of Ancient Man. An Essay on Speculative Thought in the Ancient Near East* (Baltimore: Penguin, 1949), p. 61.

9.   Here I use the masculine pronoun because the ancient Egyptians refer to this god as a male. However, as you will see, Atum also has feminine attributes.

10.   See Henri Frankfort, et al., *Before Philosophy*, p. 61.

11.   Ibid., p. 23. *The Pyramid Texts* record a variety of extracts from different sources and were discovered inscribed on the walls of the chambers within the pyramid of King Unas, c. 2400 B.C.E. Inscriptions in other pyramids have been discovered. Together they are commonly called *The Pyramid Texts*.

12.   Ibid., p. 63.

13.   Normandi Ellis, *Awakening Osiris: A New Translation of The Egyptian Book of the Dead* (Grand Rapids: Phanes Publishing, 1988), p. 21.

14.   According to Qabala, we see this duality reflected in the so-called seven double letters of the Hebrew alephbayt: bayt/vayt, ghimel/djimel, dallet/thallet, khaf/kaf, phay/pay, raysh/raysh(soft), and tav/thav.

15.   See Carlo Suarès (1), *The Cipher of Genesis: The Original Code of the Qabala as Applied to the Scriptures* (Acton, CA: Shambala, 1970).

16.   I have put this collection of nine souls together using four sources primarily: See Sir E. A. Wallis-Budge (2), *Egyptian Religion: Egyptian Ideas of the Future Life* (London: Arkana, 1987), pp. 163–68; Normandi Ellis, *Awakening Osiris*, pp. 23–24; Lucie Lamy, *Egyptian Mysteries* (New York: Thames & Hudson, 1981, 1989), pp. 24–27; and Gerard and Betty Schueler, *Coming Into the Light: Rituals of Egyptian Magick* (St. Paul: LLewellyn, 1989), pp. 26–28.

17.   Although I refer to this energy throughout the book, we won't really discuss it until chapter 6. For now, ch'i is a mysterious energy which is also known in Buddhism as the void out of which everything arises, including our thoughts and feelings.

18.   Henri Frankfort et al., *Before Philosophy*, p. 21.

19.   Normandi Ellis, *Awakening Osiris*, p. 43.

20.   Hakim claims the Egyptians named this temple this way because they believed there is both timelessness and time. It's found in the relation of the temple's orientation to east and west. As the sun rises and passes overhead, sunlight passes

through the temple, reminding the priests inside of the passage of time. Night symbolizes timelessness when there's no sunbeam. The temple was built during the 19th dynasty (1300–1196 B.C.E.) during the reign of Ramses I. That's why it's called the Ramesseum.

21.   There are many. Take aakhu, sometimes written as akh. Akh meant the spirit. Compare this with kha which means the body.

22.   Dirac's equation is a mathematical expression invented by physicist Paul Dirac in order to explain the behavior of electrons moving very near to the speed of light. Dirac discovered that all particles of matter move at the speed of light following jagged paths through space. This "jitterbugging" motion produces the illusion that matter is moving more slowly than light. He also showed that every subatomic particle is capable of existing below the threshold of perception and that an infinite number of like particles must exist at that level. When certain energies are created, one of these particles can be made to manifest out of nothing, leaving behind a hole. This hole also has physical properties and appears as the antiparticle of the particle that manifests.

23.   These patterns either manifest as particles and antiparticles, or they act as winds buffeting the manifested particles about. No matter how it may wish to avoid it, every particle of matter is continually bombarded by the incessant and hidden energy processes of the vacuum causing it to jiggle and perhaps follow occult orders whose causes are not to be found in the material world. This turns out to be a clue to the way the soul and the body communicate.

24.   See Fred Alan Wolf (8), *The Eagle's Quest: A Physicist's Search for Truth in the Heart of the Shamanic World* (New York: Summit, 1991) for more about my first meeting with Suarès.

25.   See Carlo Suarès(1), *The Cipher of Genesis*; Carlo Suarès (2), *Les Spectrogrammes de l'Alphabet Hebraïque* (Geneva: Mont-Blanc, 1973); and Carlos Suarès (3), "The Cipher of Genesis," *Tree 2: Yetzirah*, from a lecture by Suarès, reprint from *Systematics* 8, no. 2 (September 1970), edited by David Meltzer (Santa Barbara: Christopher Books, 1971).

26.   In mathematics, George Cantor created the concept of different orders of infinity which he symbolized by the letter aleph. Some infinities are "bigger" than others. Aleph-one is a larger infinity than aleph-naught and is the number of points on a line, any line. Aleph-naught is the number of cardinal numbers, the ones that go one, two, three, etc. There are an infinite number of cardinals. But there are more points on the tiniest line than numbers. Angels on the pinhead, indeed. (There are simple proofs that aleph-1 includes aleph-0, thus it is greater.)

27.   Since the Hebrew letter raysh (r-y-*sh*) contains sheen (which means hidden movement of cosmic possibility and the "breath of God") as a letter hidden in its spelling, one might consider that light energy contains the possibility of altering anything physical symbolized in modern physics by the quantum wave function. In fact, light waves are their own quantum wave functions. In this sense the light of God is not just a metaphor. Light is God's breath. In turn, the usual symbol for this wave function has unconsciously and surprisingly been adopted by most physicists with the Greek letter, ψ (psi), which resembles sheen in form, ש.

28.   There is also an element of the trickster in the action of qof because such an action is unpredictable. This idea will be discussed more fully in chapter 10.

CHAPTER 5
Resurrection Physics: a Lesson from the Land of Joking Smoking Skulls

1.   From Phil Cousineau, ed. *Soul: An Archaeology* (San Francisco: Harper-SanFrancisco, 1994), p.1.
2.   See Fritjof Capra and David Steindl-Rast, *Belonging to the Universe: Explorations on the Frontiers of Science and Spirituality* (San Francisco: HarperSanFrancisco, 1991); Jean Charon, *The Unknown Spirit* (London: Coventure, 1983); Francis Crick, *The Astonishing Hypothesis: The Scientific Search for the Soul* (New York: Charles Scribner's Sons, 1994); Paul Davies, *The Mind of God: The Scientific Basis for a Rational World* (New York: Touchstone, 1993); Herbert Van Erkelens, "Wolfgang Pauli and the Spirit of Matter," in Pakka Lahti and Peter Mittelstadt, eds., *Symposium on the Foundations of Modern Physics 1990: Quantum Theory of Measurement and Related Philosophical Problems* (Teaneck, NJ: World Scientific Publishing, 1991); Roger Penrose, *Shadows of the Mind: A Search for the Missing Science of Consciousness* (New York: Oxford University Press, 1994); Frank J. Tipler, *The Physics of Immortality: Modern Cosmology, God and the Resurrection of the Dead* (New York: Doubleday, 1994); and Danah Zohar, *The Quantum Self: Human Nature and Consciousness Defined by the New Physics* (New York: Quill/William Morrow, 1990).
3.   Tipler, *The Physics of Immortality*, p. 1.
4.   A light-year, although it sounds like a period of time, is the distance that light covers in traveling for one year. Since light moves at roughly 186,000 miles a second, a light-year is about 65.5 billion, billion miles. Our galaxy, the Milky Way, is about 125,000 light-years in diameter and around 6,000 light-years thick.
5.   The model most accepted by cosmologists is called the Friedmann model. See Charles W. Misner, Kip S. Thorne, and John Archibald Wheeler, *Gravitation* (San Francisco: W. H. Freeman, 1973), pp. 733–41.
6.   Of course this is just a model based on a number of observations, and, most importantly, a consistent theory that adequately solves problems. At the time of the printing of this paperback edition of *The Spiritual Universe*, new astronomical evidence appears to convince some scientists that the universe will go on expanding forever. However, the situation is by no means settled. My model of the soul is consistently based on an expanding and then contracting universe.
7.   Charles W. Misner et al, *Gravitation*, p. 719.
8.   It is easy to get confused here. The universe is pictured as the surface of an expanding balloon. But the "radius" of that balloon is *not* the physical radius of the universe even though cosmologists refer to it that way. The universe really exists in four dimensions (the fourth being time), not three. The space of the universe is itself "curved" in these dimensions and impossible to visualize. By reducing the three-dimensional volume of the universe to a two-dimensional surface of a balloon we can visualize what is meant by expansion and contraction of a "curved" universe. Hence in three-dimensional space a light wave could really expand outward on a three-dimensional spherical surface. But in a three-dimensional universe reduced by a dimension to the surface of a sphere, the spherical light-wave, in turn, must be reduced to a circle.
9.   A *micron* is a millionth of a meter. A *nanosecond* is a billionth of a second.
10.   This may seem confusing. Mathematicians call this *asymptotically* approaching a finite end. This means you keep getting closer but you never reach the

end. With all processing speeding up, what took one nanosecond will take only one tenth that in the next nanosecond and one tenth of that in the next. So in the last nanosecond of existence, an infinite amount of relative time will pass.

11.   Some of you might be wondering about my calculations. You must consider the fact that in this model light and the 90 percent light-speed expanding circle of technology labeled "life" move on the surface of a sphere (the universe) which itself is expanding. Thus something moving at 90 percent light-speed will not have traveled 90 percent the distance that light has traveled in the same time because of the expansion of space. In brief, during the expansion phase, the farther something is from the earth the faster it is moving away from the earth and the greater is the stretch of the spatial fabric. Hence light gets a boost from the greater stretched spatial fabric of the universe relative to the slower moving technosphere behind it.

12.   Jeremy Hayward and Francisco Varela, *Gentle Bridges: Conversations with the Dalai Lama on the Sciences of Mind* (Boston: Shambhala, 1992), pp. 152–53.

CHAPTER 6
Mind, Soul, and Zero-Point Energy

1.   It's hard to know just when it will come to a crashing end according to different omega point theories. Tipler's estimates are based on additional assumptions; namely, that there is so much future still in store we can use the future, even more than the past, to provide a better guide for the present. Tipler's model holds that the laws of physics should permit life to go on forever, until the end of time. Other cosmologists might not agree. Most estimates of the universe's lifetime are based on observations of the amount of matter in the universe, its relative size, and something called the Hubble time which provides a measure of how quickly the universe is expanding. The Hubble time is between 5 and 20 billion years. Depending on its value, one could get different estimates. Another factor is the amount of mass in the universe. Too little, and the universe will go on expanding forever, too much and it will contract much sooner. See Frank J. Tipler, *The Physics of Immortality: Modern Cosmology, God and the Resurrection of the Dead* (New York: Doubleday, 1994), p. 62.

2.   A term used by physicist John A. Wheeler.

3.   The *uncertainty principle* or *principle of indeterminism*, as it is often referred to, is the heart of the mystery we call quantum mechanics and states that it is impossible simultaneously to determine, or measure, or become aware of, by any physical means, both the position and the momentum of any material object. It is still misunderstood by many people, even a few physicists. It reflects the inability to predict the future based on the past or based on the present, and arose from the ideas and thoughts first stated by Werner Heisenberg around 1926 or 1927. The Uncertainty Principle provides an understanding of why the world is made of events which cannot be related entirely in terms of cause and effect and is at the root of all physical matter interactions which may manifest as doubt and insecurity in human beings. If this is so, once it is fully understood it could create a condition of enlightenment in which the world is seen as an illusion and as a product of mind or consciousness. Thus, every corner of empty space potentially has particles present, since no one of them can be located precisely.

4.   Temperature is usually measured on a Fahrenheit or Celsius (Centigrade) scale. You probably remember the formula, F = (9/5)C + 32, which says that the Fahrenheit temperature, F, is computed by multiplying the Centigrade temperature, C, by 1.8 and then adding 32. What you are doing is changing scales so, for example, water boils at 212 degrees and freezes at 32 degrees Fahrenheit while the Celsius scale fixes temperature arbitrarily so that water boils at 100 degrees and freezes at 0 degrees Centigrade. The absolute scale was taken from Swedish astronomer Anders Celsius's notion of degrees, namely, that 100 degrees separates boiling from freezing water, but adjusted the zero downward by a 273.3 degrees, where, supposedly, everything freezes and nothing moves, not even an atom. But, temperature is a classical physics-based concept and does not take into account the quantum-physical zero-point motion.

5.   This means mass-energy equivalence as in the famous formula of Einstein, $E = mc^2$.

6.   The sun has about 2×10^{33} grams of matter. If it is radiating away 5×10^{12} grams of mass each second you might wonder how long the sun could last. At this rate it will last 400 billion, billion seconds or about 12,684 billion years. However, it will bellow out as a red giant star in around 7 billion years, so don't worry.

7.   According to Einstein's equation, $E = mc^2$, energy changes to mass and back again.

8.   See the above note.

9.   Coherence is very important, and I don't have enough space to deal with it here. It essentially is a property of quantum physical wave fields—their ability to hold things together by not specifying exactly where they are.

10.   The last discovery was the top quark, a tiny but quite massive bit of matter (on the scale of subatomic particles). Quarks combine to make nuclei of atoms which have much smaller masses than quarks have. The missing mass is turned into energy-glue holding the quarks together. Ultimately even quarks are subjected to quantum clouds of possibility.

11.   Just as an electron has mass and electrical charge, so does the positron, the antimatter partner of the electron. It has exactly the same mass as the electron, but the opposite sign of electrical charge.

12.   The idea here is that ordinary matter, seemingly stable, really isn't and is just a longer lived fluctuation of the vacuum zero-point energy.

13.   Difficult, but not impossible. This field of study, called *quantum electrodynamics*, has been quite successful, although there are still some difficulties dealing with particle self-energies.

14.   Antiparticles are particles of antimatter, as you may have already surmised. See Richard P. Feynman and Steven Weinberg, *Elementary Particles and the Laws of Physics: The 1986 Dirac Memorial Lectures* (New York: Cambridge University Press, 1987), pp. 1, 2.

15.   The electron is the smallest subatomic particle. It has certain measurable properties. These include electrical charge, inertial mass or resistance to accelerated motion, spin (which can be thought of roughly by picturing the electron as a tiny ball spinning about an axis), and electron exclusion (the tendency to avoid another electron by not entering the same quantum physical state, which appears whenever two or more electrons are near each other). The photon is the smallest unit of light energy. It has measurable properties consisting of no electrical charge, no inertial mass—

although it is capable of delivering a "punch" of momentum and has a spin twice the magnitude of an electron's. The proton carries an electrical charge equal (but opposite in sign) to the electron and has a spin equal to it, but a mass 1,836 times heavier. Compared with quarks, which compose protons, the proton is infinitely stable. No one has ever seen one decay.

16.   Antimatter, positrons, can be created in our universe during certain radioactive decays of specific atomic nuclei. Whenever an electron and a positron combine, that is, when they annihilate each other in the process, they give off gamma radiation—high-frequency light waves. The positive energy of the electron, mc^2, is $2mc^2$ above the negative energy, $-mc^2$, of the positron—the hole in the Dirac Sea. Hence, $2mc^2$ of energy is released in the matter/antimatter electron/positron annihilation. This process is commonly used in hospitals everywhere to diagnose illness. By injecting radioactive isotopes into the blood stream and then observing the gamma rays emitted, a positron-emission-tomography or PET scan is obtained revealing, for example, the location of a cancerous tumor.

17.   I only mention ch'i in passing here. Ahead in chapter 10 I'll have more to say about ch'i and its interplay with the soul.

18.   See John Davidson, *The Secret of the Creative Vacuum: Man and the Energy Dance* (Essex, England: C. W. Daniel, 1989), pp. 221–28.

19.   F. A. Wolf (2), *Star Wave: Mind, Consciousness, and Quantum Physics* (New York: Macmillan, 1984), p. 265.

20.   Jean Charon, *The Unknown Spirit* (London: Coventure, 1983), p. 65.

CHAPTER 7
Quantum Evidence: The Self and the Soul

1.   From *The Rubaiyat*, trans. A. J. Arberry (London: Emery Walker, 1949). Also see Shems Friedlander, *The Whirling Dervishes* (New York: SUNY, 1992), p. 76.

2.   Neil Douglas-Klotz, *Desert Wisdom: Sacred Middle Eastern Writings from the Goddess Through the Sufis, Translations and Commentary* (San Francisco: HarperSanFrancisco, 1995), p. 151.

3.   See Malcolm W. Browne, "The Soul Weighs In: Scientist Says He Can Measure Our Spiritual Selves," *San Francisco Chronicle: Sunday Punch*, taken from the *New York Times* (January 16, 1994).

4.   Even though I see the soul in terms of particles, I don't really believe the soul is made of real particles. One way to deal with this is to imagine that the soul is composed of *virtual* particles—those that exist with negative energies beneath the Dirac Sea, discussed in the previous chapter. I do believe that these virtual particles composing the soul communicate with the real particles that make up our body. Spin plays an important role in this communication, and here we see that role.

5.   If the virtual particle is an electron, the duplicate particle is a real electron, for example.

6.   As I explained in the previous chapter, if the virtual particle is an electron, when it "pops" out into the real world and appears as a real electron, the hole it leaves behind is the antimatter partner—the positron.

7.   In case you wondered, the reference direction of the detection apparatus is defined by the direction of an internal magnetic field.

8.   This means they undergo some form of collision wherein they are moving relatively slowly toward each other and, consequently, do not come too close to each other. You may think of this as a glancing collision.

9.   Also known as the *parallel worlds* or *parallel universes* model. I want to hold it to just mind-possibilities. I don't need universes here.

10.   Actually, I'm fudging a bit here. As I pointed out, since no particular direction has been defined until someone attempts to measure a direction, the two particles spin in opposite directions in every direction simultaneously! With respect to any specific direction, however, just two mind-possibilities are open. One could imagine God as a magician saying "pick a direction, any direction," as She or He produces a deck with an infinite number of cards in it. As soon as you think of a direction, the deck reduces to just two cards. Or you could imagine God with the infinite deck asking you to pick any two cards at random. No matter what you choose, the two cards will always be opposite pairs like one black and one red queen or one black and one red six.

11.   Erwin Schrödinger, the inventor of the equation that describes the wavelike behavior of the probability wave function, referred to this situation as an entanglement. Particle A is so entangled with B that it no longer sees itself as separate. However, Schrödinger would not have given observational power to either particle. Positing the particle with the ability to observe is my conjecture.

12.   The word *remembering* implies memory, not necessarily consciousness at this point. Although this is at present a moot point, soon I shall make a distinction between remembering and conscious awareness.

13.   Since one of the pair remains a virtual particle, they are not physically separated. Remember, we are considering the interaction of a particle with a virtual mate, not a hole which would be an antimatter partner. In that case, the particles would necessarily be physically separated.

14.   Thus since A is not separated from B physically, A is in interaction with everything and nothing! That is the paradox of interacting with the vacuum.

15.   Because they are identical particles, their entanglement is far greater than it would be if they were not. So long as no observation of either particle occurs, the entanglement lasts. Since all electrons are identical in every respect with one another, and all protons, neutrons, and atoms are also, we surmise that we humans, in spite of our great differences, are identical to each other at some level of interaction since we are made of identical particles.

16.   As far as spin and identity were concerned they would be independent of each other. However, they would still feel the electrical force of each other's charge. It turns out that the vanishing of one of the two mind-possibilities in the parallel worlds model of quantum physics expressed here is equivalent to the action of consciousness in the collapsing probability model expressed in chapter 12. I'll go into the consciousness-collapsing model in more detail there.

17.   This action is complementary in the quantum physical sense to the previous over-soul (spin-0) observation which led to nonseparation.

18.   Bertrand Russell spells this out quite clearly: ". . . the Cartesian dualism presents two parallel but independent worlds, that of the mind and that of matter, each of which can be studied without reference to the other. . . . the mind does not move the body [and] . . . the body does not move the mind. . . . In the whole theory

of the material world, Cartesianism was rigidly deterministic. Living organisms, just as much as dead matter, were governed by the laws of physics; there was no longer need of [a] soul." Bertrand Russell, *History of Western Philosophy and its Connection with Political and Social Circumstances from the Earliest Times to the Present Day* (London: George Allen & Unwin, 1961), p. 551.

19.   I discussed this rather fully in Fred Alan Wolf (10), *The Dreaming Universe: A Mind-Expanding Journey into the Realm Where Psyche and Physics Meet* (New York: Touchstone, 1995). I have more to say about it in what follows.

20.   By extraction, I mean when something outside the system attempts to gain knowledge about the system. This is the normal way information-exchange takes place. One says, "I see the apple or I see the electron has *spin-up*." The knowledge of the apple is extracted from the apple. The knowledge of the spin is extracted from the spinning particle. Something about the apple and the spin is defined. I call this *knowledge entering consciousness.*

21.   See David Z. Albert, "Self-Measurement," chap. 8 in *Quantum Mechanics and Experience* (Cambridge: Harvard University Press, 1992), pp. 180–89. I must point out that I am giving an essentially original, somewhat colorful, and perhaps oversimplified interpretation of the process of self-measurement, and Albert may not agree with my interpretation. Nevertheless, the essential point of this indicates that self-measurement violates the uncertainty principle, something I believe he would agree with since he was the first to realize it. You can know things about yourself that you can never even hope to learn about another.

22.   By outside world I mean another physical system, say particle C, an interested bystander.

23.   Yes. Salvation is at hand. This knowledge content or memory of A can be symbolized by writing it as $[0, 1]$. The big zero symbolizes that A has the spin-0 awareness and the 1 symbolizes that it simultaneously has knowledge of its own identity—A knows in which direction its spin is pointing. This quantum piece of evidence associated with quantum-mechanical computer automata produces the difference between self and the outside world.

24.   In symbolic depiction: $[0, 1] \rightarrow [1]$ & $[?]$. The big question mark symbolizes the loss of soul knowledge.

25.   And the analysis would become even more complex. According to Leonard I. Schiff's *Quantum Mechanics*, 3rd ed. (New York: McGraw-Hill, 1968), three spin-½ particles can never exist in a spin-0 state. Instead they take on values: 1½ or ½ . It turns out that there are four ways for them to have the total value 1½ and four ways (actually divided into two groups) that they can take on the value ½. Nevertheless, the three particles would behave as a single system in much the same manner as I described for the two spin½ particles.

26.   System-C can acquire that knowledge without ever knowing, for example, in which direction A's spin is pointing. However, if it asks the direction of A's spin then it too will fall into A's world and lose knowledge of the spin-0 system.

27.   We can symbolize this communion as $[1, [0,1]]$. The bold brackets symbolize that C, the first 1 in the bracket, has joined with the original A-B pair, in the inside brackets, and acknowledged that the spin-0 state exists along with the presence of an identity A. But the smaller brackets surrounding the 0 and the 1, symbolize that C knows not the state of A. Now C is connected to the entity A-B, but not with B alone nor with A alone.

28.   This can be symbolized by [1, [0,1]] → C[1] & A[1] & [?]. Now C and A each have separate self-knowledge, but no knowledge of each other nor of their communion with B.

29.   Adam is in the [0, 1] state of bliss.

30.   The state of them all is [Δ, 1], where the triangle symbolizes the trinity that each knows secretly and collectively, while the 1 symbolizes the id-entity that each knows secretly and separately.

31.   To call God by the plural *They* may seem even weirder than any of the ideas I put forward here. But in the very first sentence of the Hebrew Bible, God is referred to as Elohim, the feminine plural form of the Hebrew word for *Eloh* or *Allah*.

32.   In symbolic form: [Δ, 1] → A[1] & E[1] & [?]. We might refer to this as the present condition of humanity after the expulsion from the *garden*. Each knows only itself and feels no connection with the others.

33.   David Z. Albert dealt with this issue in Albert (2) "On Quantum-Mechanical Automata," *Physics Letters* 98A, nos. 5, 6 (October 24, 1983), pp. 249–52. There he said: "There is something subjective . . . about the capacity of a [memory device to make predictions] because that capacity depends . . . on its *identity*. There are some combinations of facts that can in principle be predicted by an automaton *only about itself.*" Thus, Albert was the first to recognize that self-measurement has an essentially different character from other-measurement, namely: "If a quantum-mechanical automaton were . . . to look at itself and measure itself and produce a description of itself, that description would be different, . . . in *nature*, from any description it might produce of an external object. . . . [One might ask] whether such an automaton *might* be a model of our *own* . . . experience."

CHAPTER 8
The Buddhist Nonsoul

1.   Quoted in Huston Smith, *The Religions of Man* (New York: Harper and Row, 1958), p.129.

2.   Einstein was similar in his denial of the extraordinary aspects of quantum physics. His famous denial proof indicating that something was missing from quantum mechanics, known as the Einstein-Podolsky-Rosen experiment, became the cornerstone of quantum mechanics. This proof showed that its nonordinary character was absolutely *essential*. See my discussion of this in Fred Alan Wolf, *Taking the Quantum Leap: The New Physics for Nonscientists*, rev. ed. (New York: HarperCollins, 1989).

3.   Nancy Wilson Ross, *Buddhism: A Way of Life and Thought* (New York: Alfred A. Knopf, 1980), p. 31.

4.   Consequently, here I shall concentrate on the first axiom: the meaning of dukkha. In chapter 9 we will look at the second axiom: how dukkha arises—the detailed mechanics of illusion. Chapter 10 deals with the tricksterlike consequences of the illusion. In chapter 11, I return to the third axiom (how dukkha ceases) and the fourth axiom (the path leading to dukkha cessation).

5.   The both/and aspect of this lifestyle is very important as I shall explain ahead and in greater detail in chapter 12.

6.   In case you forgot your thermodynamics, the second law states that entropy—the measure of disorder in a physical system—will generally increase when

things change. This means that no matter how hard we try, there is always a little more chaos in the universe after we do something, always a little energy lost to friction, for example, after we start our engines, turn on the TV, or take a bath. It basically says any system left to its own devices will degrade order to make chaos, if no work is done to compensate for the degradation.

7.   Such logic, when reduced, leads to the modern computer which consists of elements that are either 0 or 1, but never anything in between.

8.   A *koan* is an unanswerable question usually put to a student by his or her *roshi* (master). Some famous koans include, "What was the face you had before your grandparents were born?" and "What is the sound of one hand clapping?"

9.   This superposition principle means holding both definitions in mind simultaneously, something that at first seems impossible. Actually we unconsciously do this all the time and hardly notice it. It's called *chilling out* in the vernacular. The Buddhists consciously develop this skill to a fine art. You can too, but it takes some practice. I'll have more about the both/and aspects of quantum physics in chapter 12.

10.   Walpola Rahula, *What the Buddha Taught*, 2nd ed. (New York: Grove Press, 1974), p. 62.

11.   I'm not exactly sure what this theory was. I can guess that it meant the existence of an everlasting entity—a soul that eventually joins with Brahma at some point in its evolution. Also, in case you are wondering, a *brahmana* is simply a Brahmin, a member of the highest, priestly, caste in India.

12.   It would appear from this remark that Buddhist theology is not annihilationistic either. However, I believe this theory posits the existence of a soul that persists for some time and then is annihilated, presumably at death.

13.   Here Rahula, *What the Buddha Taught*, in a footnote on p.63, insists *anatta* means without a self. I have assumed that what the Buddha called the self is what he meant by the soul or *Atman*. For the Buddha also knew conventional wisdom and knew that in ordinary usage one refers to oneself as a convenience of thought and speech. The Buddha refuted all concepts of soul, self, and *Atman*, taking them all as illusions.

14.   J. Robert Oppenheimer, *Science and the Human Understanding* (New York: Simon & Schuster, 1966), p. 69.

15.   This quote comes from a four-part program *Death: The Trip of a Lifetime* shown on Public Television.

16.   Rahula, *What the Buddha Taught*, p. 24.

17.   I have rephrased his statements in simpler language, but since these are his paraphrased thoughts I have set off his words as if they were a quotation.

18.   Rahula, *What the Buddha Taught*, pp. 16–28.

19.   I don't believe they are identical. The goal of this book is to clarify their differences. In part 4, I explain how different they are.

20.   Ross, *Buddhism*, p.141.

21.   Ross, *Buddhism*, p. 145.

22.   John Daishin Buksbazen, *To Forget the Self*, vol. 3, Zen Writings Series (Los Angeles: Zen Center of Los Angeles, n. d.), pp. 1–3.

23.   Ibid., p. 5.

24.   Minor White quoted in Ibid., p. 7.

25.   Ibid., p. 7.

26.   I can't stress how important this is. Without reflection, there is no universe. More about this in part 4.

27.   Quoted in Buksbazen, *To Forget the Self,* p. 19.

28.   Quoted in Ibid., p. 27.

29.   Ross, *Buddhism,* p. 168.

CHAPTER 9

Good, Evil, and Soul Addiction

1.   Frieda Fordham, *An Introduction to Jungian Psychology* (New York: Penguin, 1953), p. 50.

2.   Remember that spinning virtual electrons, those with $E = -mc^2$ energy in the Dirac Sea, are in continual interaction with spinning real electrons, those with $E = +mc^2$ energy, as I described in chapters 6 and 7.

3.   The battlefield of time and timelessness occurs everywhere in the universe, in our minds and bodies, in nature, in plant and animal life, in the formation of stars and planets, and in the creation of the universe itself! The battle takes place very rapidly on a tiny stage indeed. The scale of time is called the Planck time and equals 5.391×10^{-44} seconds. The spatial scale, called the Planck length, equals 1.616×10^{-33} centimeters. On this space-time scale, space, time, and matter seemingly interchange and bubbles of nothingness manifest out of space and time, which in turn fluctuates wildly. Physicist John A. Wheeler calls this *quantum foam.*

4.   There is certainly a lot more to this battle. Somehow resistance must arise along with time in order to produce what seems to us to be a unidirection to our experience of time.

5.   As I mentioned in a endnote to chapter 5, the issue of the eventual end of the universe is not closed. I believe, like Dirac, the beauty of the model will ultimately decide the fate of the universe! This is certainly a strange belief but, based on the work of some theoretical physicists whose discoveries were based on beauty rather than logic, a reasonable one.

6.   Not only is sight involved, but all of the six Buddhist senses. More about this later.

7.   I return to this in the final chapter where much of this is summarized.

8.   Simply put, if the soul is the volume of a sphere, for example, the ego would be its outer skin, its surface.

9.   See chapter 10.

10.   My own experience with cigarette smoking illustrates this. I gave up the habit soon after I realized that smoking made no sense and that I was smoking to fulfill an image of myself which was tied to smoking heroes I saw in popular movies when I was growing up. I was quite fond of watching Alan Ladd and the way he smoked. I tried to emulate him. In a sense, I carried him for many years of my life as the monk carried the image of the girl. When I no longer believed I was this glamorous image, the desire for smoking began to abate.

11.   Mark Epstein, "Are We All Hungry Ghosts?" *Inquiring Mind,* 11, no. 2 (spring 1995), p. 11.

12.   I use this term in the Jungian sense to mean "meaningfully related other than through cause and effect." Such events usually happen at the same time, but as

I have learned from Jung's disciple, Marie Louise von Franz, they need not be simultaneous to be synchronistic.

13.   Walpola Rahula, *What the Buddha Taught*, 2nd ed. (New York: Grove Press, 1974). Also see Nancy Wilson Ross, *Buddhism: A Way of Life and Thought* (New York: Alfred A. Knopf, 1980).

14.   Born in 1707, Leonhard Euler was a Swiss mathematician. He wrote *Introduction to the Analysis of Infinity* in 1748 and taught in Berlin and St. Petersburg. Euler represented class relationships by spatial relations called Euler Diagrams. Here I use Euler circles to represent a syllogism.

15.   The smallest bits seem to last the longest. The proton and the electron appear to be absolutely stable; however, even they are continually rocked by the forces of the vacuum.

16.   David Bohm, the noted physicist and philosopher, realized this in many of his writings. He began to believe that somehow the electron could back-react and thus alter the wave. But so far no one knows how to calculate this in a mathematically consistent way.

17.   See Fred Alan Wolf (8), *The Eagle's Quest: A Physicist's Search for Truth in the Heart of the Shamanic World* (New York: Summit, 1991).

18.   Amit Goswami makes this a major point in his book. See Amit Goswami, *The Self-Aware Universe: How Consciousness Creates the Material World* (New York: Tarcher/Putnam, 1993).

19.   I use the word *consciousness* to describe a broader concept than *awareness*. *Awareness* I take to mean specific experiences of one's senses. I use consciousness to include these experiences as special cases of something much harder to define. For example, I hope that you are conscious of your soul even though you may not be aware of it. I am conscious of compassion even though I am not aware of any feelings of compassion arising in my body as I write these words.

20.   I tell you more about these organs of sense in chapter 11. Meanwhile there are six: eye and sight, tongue and taste, skin and touch, ear and sound, nose and smell, and mind and thought.

21.   This is a tricky point. One would certainly assume that there are many interactions occurring in the universe of which we are only aware of a few. The question is can any of them exist without some entity sensing them? Buddha would argue that since there is no evidence to indicate they could, they don't.

22.   E. J. Gold, *American Book of the Dead* (San Francisco: AND/OR Press, 1975), pp. 7–12.

23.   Ibid., p. 12.

CHAPTER 10
The Soul Trickster: Chaos, Lies, and Order

1.   This quote was paraphrased by Richard Smoley in his article, "My Mind Plays Tricks on Me," *Gnosis Magazine* (spring 1991), p. 12, and is attributed to Boris Mouravieff, known as "the mysterious Russian sage."

2.   Karl Kerényi, *Hermes: Guide of Souls*, trans. Murray Stein (Dallas: Spring Publications, 1992).

3.   Ibid., p. 53.

4.   Some would argue the trickster is actually number one, as the first icon, belonging to the fool, has the number zero. However it is known that the twenty-two cards of the major arcana of Tarot are associated with the twenty-two symbols of the Hebrew alphabet. The Hebrew character aleph, representing the number one, appears as the first symbol of both the Qabala and the symbol connected with the fool in the Tarot. This seems to be a stronger indication that the first card of the arcana begins with the number one and not zero, so the trickster/magician is connected with the number two.

5.   Smoley, "My Mind Plays Tricks."

6.   Albert Einstein, "Geometry and Experience," [a lecture to the Prussian Academy of Sciences, January 27, 1921] in *Ideas and Opinions* (New York: Crown, 1954).

7.   Fritjof Capra, *The Tao of Physics* (Boston: Shambhala, 1975), pp. 213–14.

8.   See chapter 5 and figures 5.2 through 5.5.

9.   I took this from a television interview with her on Public Television. However, I don't remember the source. Some of her work is also described in Michael Talbot's book, *The Holographic Universe* (New York: HarperCollins, 1991), p. 112 and in Larry Dossey's book, *Recovering The Soul: A Scientific and Spiritual Search* (New York: Bantam, 1989), pp. 85–86.

Chapter 11
Heaven, Hell, Immortality, Reincarnation, and Karma

1.   Carlo Suarès (2), *Les Spectrogrammes de l'Alphabet Hebraïque* (Geneva: Mont-Blanc, 1973), p. 31.

2.   Karen Armstrong, *A History of God: The 4,000-Year Quest of Judaism, Christianity, and Islam* (New York: Alfred A. Knopf, 1994), p. 290.

3.   See chapter 8. Bodhidharma was a far-wandering iconoclastic and apocryphal Indian Buddhist master.

4.   Alexandra David-Neel, *Buddhism: Its Doctrines and Its Methods* (New York: Avon, 1979), p. 202.

5.   Dannion Brinkley (with Paul Perry), *Saved by the Light: The True Story of a Man Who Died Twice and the Profound Revelations He Received* (New York: Villard Books. 1994), pp. 3–28.

6.   Remember that *transit* refers to the time spent between death and rebirth.

7.   *Lightlike* refers to the interval in space and time between two events where the spatial distance between the events equals the speed of light multiplied by the time interval separating the events. The separated events are said to be *timelike* when *anything* material can connect one event with the other. Timelike events can always be cause-and-effect related. The events are said to be *spacelike* separated when *nothing* material can connect one event with the other. This is because the events are so far apart that the material object would need to travel at a speed greater than light-speed, which is impossible according to relativity. Presumably, transit occurs along spacelike trajectories, those that mark a path through space and time of a nonmaterial object moving faster than light.

8.   William P. Alston, ed., *Religious Belief and Philosophical Thought* (New York: Harcourt Brace Jovanovich, 1963), p. 340.

9.   The spelling of this word varies depending on the source. It is also spelled khanda or khandha.

10.   This is difficult to explain as any Buddhist will tell you. I use *implication* to suggest the thing exists and is called the soul. However, I want to couch this in Buddhist language as much as possible. So for the moment take it that whatever it was can be conveniently thought about as a soul without necessarily taking for granted that a soul actually is present. Oh, well, I tried to be clear.

11.   Hirabhai Thakkar, *Theory of Karma* (Amreli, India: Shri Rameshwer Printing, 1988), p. 8.

12.   For a complete look at how we use light, X-rays, magnetic resonance, and sonar to see into the body, see Fred Alan Wolf (4), *The Body Quantum: The New Physics of Body, Mind, and Health* (New York: Macmillan, 1986).

13.   In relativity theory, time is often referred to as the "fourth" dimension. In the equations of relativity, if you multiply the symbol representing time, t, by i, the square root of minus one, the symbol it appears as if it were a real spatial dimension like the other three, although it is purely an imaginary number.

14.   I refer the reader to chapter 5 and the comments of the Dalai Lama about computers being able to cognate.

15.   I mean *seen* metaphorically. The soul senses all six types of willful action and thus can *see* mental formations. However, it may not be able to differentiate between the six forms of willful action or karma posed by Buddhist thought. I'll leave this ability now for my next book.

16.   Y. Aharonov and M. Vardi, "Meaning of an Individual 'Feynman Path'," *Physical Review Digest*, 21, no. 8, (April 15, 1980), pp. 2235–40.

17.   A quantum system is any physical system that is simple enough for its quantum physical state to be observable. Subatomic matter in the form of particles are certainly quantum systems. Larger aggregates of matter could be quantum systems but usually don't qualify because far too many transitions are occurring in them to be observed by an outside observer. Matter in our brains may be composed of simpler quantum systems.

18.   Some of my readers may recognize the quantum watched pot from my previous book, *The Dreaming Universe: A Mind-Expanding Journey into the Realm Where Psyche and Physics Meet* (New York: Touchstone, 1995). All of this might be considered to be just quantum physics speculation. However, in 1989, physicist Wayne Itano and his colleagues at the National Institute of Standards and Technology in Boulder, Colorado, actually experimentally observed the "quantum watched pot" and, indeed, it never boiled! Their experiment involved watching some 5,000 beryllium atoms confined in a magnetic field and then exposed to radio waves of energy. The atoms were the equivalent of the pot of water and the radio waves the equivalent of the heat applied to the pot. Under such circumstances the atoms will "evolve" into excited atomic energy states as they absorb the radio energy. Nearly all 5,000 will reach their excited state goals in a little over 250 milliseconds (ms), that is, a quarter of a second.

To check this the physicists would observe the atoms after 250 ms by shining a short pulse of laser light into their midst. Excited atoms do not absorb and immediately reemit the laser energy. Atoms that remained in the unexcited state do. So, by observing the scattered laser light after it passed through the trapped atoms, the physicists were able to determine just how many atoms remained unexcited.

Virtually none were after 250 ms. We could refer to this as the unwatched pot that naturally evolved to the boiling state in a quarter of a second. But then the scientists became slightly vigilant. They decided to look at the atoms halfway along, after 125 ms had passed. So an eighth of a second after starting the experiment the laser pulse was turned on and then at the 250 ms mark the scientists looked again and found that only one-half of the atoms were excited. They repeated the experiment by looking in at 62.5 ms, 125 ms, 187.5 ms, and 250 ms; in other words, they divided the one-quarter second interval into four equal parts. They were surprised to find that their enhanced vigilance resulted in only one-third of the atoms making it to the excited energy state by the end of the complete period of 250 ms.

Next they redoubled their vigilance by looking in 16 times, 32 times, and finally 64 times during the 250 ms interval. In the final experiment where they watched their tiny atomic "pots" in 64 equally spaced tiny time intervals, virtually none of the atoms were ever found in an excited state, even though 250 ms had passed. They all remained frozen in their ground or original states just as they were when the experiment began. In each experiment, mind you, the "heat" was on—the radio waves were continuously sent in to the magnetically trapped beryllium atoms.

19.   Physicists Paul Kwiat, Harald Weinfurter, and Anton Zeilinger working at the University of Innsbruck in Austria have used the quantum Zeno effect referred to here to obtain information about a system that was never observed. By not looking at something they were able to obtain as much information about the system as they would have obtained by looking. See Paul Kwiat, Harald Weinfurter, and Anton Zeilinger, "Quantum Seeing in the Dark," *Scientific American* (November 1996), pp. 72–78.

CHAPTER 12
How Many Souls Are There?

1.   From Nancy Wilson Ross, *Buddhism: A Way of Life and Thought* (New York: Alfred A. Knopf, 1980), p. 48.

2.   A mathematically real quantity is a positive or negative number or any function of one. Imaginary numbers also play a role in physics. There is no way to formulate quantum mechanics without using the ubiquitous $i = \sqrt{-1}$ the unit of imaginary numbers. See my earlier books, particularly Fred Alan Wolf (10), *The Dreaming Universe: A Mind-Expanding Journey into the Realm Where Psyche and Physics Meet* (New York: Touchstone, 1995), pp. 161–63.

3.   Ian Wilson, *The After Death Experience: The Physics of the Non-Physical* (New York: William Morrow and Co., 1987), p. 177.

4.   Bobby is not his real name.

5.   Ian Wilson, *The After Death Experience,* pp. 77–78. Again, I have not used their real names.

6.   Materialists would have it that consciousness is an epiphenomenon, one that arises from complex interactions in matter. Thus there is nothing special about them.

7.   Ludvik Bass, "The Mind of Wigner's Friend," *Hermathena: A Dublin University Review,* no. 112, (1971), p. 58.

8.   Given that there is only one consciousness or one mind and, hence, one soul, it follows that the sense that your mind is separate from another's must be false.

Thus, the little bodymind or ego, the one we count when we count heads or play musical chairs, is an illusion. The bodies are real, but the separated egos are not.

9.   Strictly speaking, the mind is not the soul; however, at least circumstantially speaking, one could never have a soul without mind. Thus, in order to know if a soul is present or not one should certainly have the *presence of mind*. In chapter 13, I'll explain why it is vital for soul to have the property of awareness, while matter does not. In other words, soul possesses mind. Hence if the mind is one, the soul's unity must follow. Logically speaking, soul implies mind. Mind implies unity. Thus, soul implies unity.

10.   This Latin phrase applies to a method of proof for a logical argument. In this method, one deduces a contradiction from the opposite of the argument's conclusion. For example, take the argument, *all crows are birds, all birds are animals,* therefore, *all crows are animals.* One would take the opposite conclusion, *no crows are animals.* Next, one returns to the first statement of the argument, *all crows are birds.* From this one would conclude, *some birds are not animals* (namely crows), which contradicts the second statement, *all birds are animals.* Hence, all crows are indeed animals.

11.   Some of you may scoff at this simple example. Yet if we regard infinity as the final number, Ω, then it turns out that Ω plus one is not greater than Ω. In fact even Ω plus Ω still equals Ω. For example, consider that there are an infinite number (Ω) of odd numbers, like 1, 3, 5, and so on and an infinite number (also Ω) of even numbers like 2, 4, 6, and so on. Suppose I add the infinite set of odd numbers to the infinite set of even numbers, will I get twice infinity? The answer is no, you still get infinity—the infinite set of numbers 1, 2, 3, and so on.

12.   Ludvik Bass, "The Mind," p. 58.

13.   See Erwin Schrödinger, *My View of the World,* originally published in German (Hamburg-Vienna: Paul Zsolnay Verlag GMBH, 1961). English edition (Cambridge, England: Cambridge University Press, 1964). Reprint (Woodbridge, CT: Ox Bow Press, 1983). See also Erwin Schrödinger, *What is Life?* and *Mind and Matter* (Cambridge, England: Cambridge University Press, 1967).

14.   Note because Robotina is a robot, her *wishes* are mechanical metaphors for some computer program in Robotina that simulates choice-making. A similar situation exists with regard to any of Robotina's simulated willful actions.

15.   We are assuming that Robotina has a well-working robotically conscious mind and that Robotina is not suffering any computerized neural network-induced delusion. If this seems droll to you, please look at my book, *The Dreaming Universe: A Mind-Expanding Journey into the Realm Where Psyche and Physics Meet* (New York: Touchstone, 1995), pp. 117–19, where I discuss how such neural nets can "suffer" from obsession, fantasy, and hallucination. In no case will there be anything in discordance like the particle's spin in one direction and Robotina's mind registering the opposite (that is, in no case will there be anything like Robotina-up and spin-down).

16.   I do not choose to use the word *unconscious* here. I wish to reserve that word for processes which could result in awareness, but don't because they haven't been integrated into a single stream of consciousness. This will be a subject for another book.

17.   Roberta could be on the moon and you could be communicating with her via radio or TV signals. How would you know what Roberta was? As far as you are

concerned, Roberta could be a machine, that is, a Robotina, in which case, no matter how much Robotina protests, you will assume that she is a cleverly-built machine.

18.   Indeed, much evidence exists indicating quantum mechanics applies in all physical situations regardless of scale. See my book *Taking the Quantum Leap: The New Physics for Nonscientists*, rev. ed. (New York: HarperCollins, 1989), chap. 15.

19.   Spoken by Amit Goswami at a seminar on consciousness. Check out his book: Amit Goswami, *The Self-Aware Universe: How Consciousness Creates the Material World* (New York: Tarcher/Putnam, 1993).

20.   The main reason for the difference comes from the interference of the both/and possibilities—a fact at the heart of the weirdness of quantum mechanics. Such interference of possibilities leads to the impossibility of some physical consequences ever occurring. Noninterfering possibilities would normally give rise to observable consequences if the state of the system was either/or. For example, a stream of particles, each particle subject to both/and interference from following two or more possible pathways directed toward a screen would consistently not arrive at certain places because of the interference of the pathway-possibilities. See my discussions of the *double-slit experiment* in my book *Taking the Quantum Leap*. See also Fred Alan Wolf (6), *Parallel Universes: The Search for Other Worlds* (New York: Simon & Schuster, 1989).

21.   I heard this at a conference. I don't know to whom the aphorism belongs.

22.   Some of you might be thinking, what about having four or five billion independent minds all acting according to 2? Where's the contradiction? The answer is, this violates 1, for it would mean no nervous system follows the laws of nature that include quantum physics. Since nervous systems are composed of atoms and molecules, then none of them would follow 1 either. This would mean that quantum physics was wrong to begin with. But, so far, quantum physics is a correct theory.

23.   R. P. Feynman (3), "The Value of Science," *Project Physics Reader: An Introduction to Physics 1, Concepts of Motion*, Authorized Interim Version (New York: Holt, Rinehart and Winston, 1968–69), p. 3.

24.   Schrödinger, *My View of the World*, p.13.

25.   Quoted in Bass, "The Mind," p. 61.

26.   Indeed, the independent bees seem run by 1, rather than by 2. Yet 2 arises and connects them all together.

CHAPTER 13
Soul-Talk

1.   Taken from C. G. Jung and W. Pauli, *The Interpretation of Nature and the Psyche. Synchronicity: An Acausal Connecting Principle* (Jung) and *The Influence of Archetypal Ideas on the Scientific Theories of Kepler* (Pauli) (NY: Bollingen Foundation, Pantheon, 1955), p. 178.

2.   Swami Vivekananda, *Jnana-Yoga* (New York: Ramakrishna-Vivekananda Center, 1982), p. 140.

3.   Taken from Phil Cousineau, ed., *Soul: An Archaeology. Readings from Socrates to Ray Charles* (San Francisco: HarperSanFrancisco, 1994), p. 202. The original is from *Metamorphoses of the Soul*, vol. 2.

4.   *Nonlocality* is a bit of quantum physics jargon. Normal experience with interacting objects (imagine holding two bar magnets in your hands) shows that if one object affects the other, the effect grows weaker as the objects are moved farther from each other. The objects are said to be *locally* connected. The simplest description of *nonlocality* is when a single event occurs at two distinct places in space or time, as opposed to two distinct events influencing one another like bar magnets. As an example of nonlocality, imagine watching a football game on many television screens simultaneously, as in a large department store. You see the same thing repeated on different screens placed at different locations in the store. If the program were taped, you could even see the same thing at different times. One could conclude, if one didn't know better, that the different games were actually occurring at different locations and many different times, played by tiny beings all living within the television sets. Somehow there was a very strong interaction between them, regardless how far apart the television sets were placed. In physics nonlocality refers to objects that strongly interact in rhythm no matter how far apart they are, as if they were one objective series of events like the football game rather than separate objective events like a hockey game and football game. What happens to one of them instantly affects the other with no change in strength, even when they are light-years apart. One says that one object is nonlocally connected to the other. It would be as if whenever a football player scored a touchdown, a hockey player would simultaneously score a goal.

5.   I originally used this example as it was told to me in Wolf (4), *The Body Quantum: The New Physics of Body, Mind, and Health* (New York: Macmillan, 1986), p. 260, as a model dealing with health and illness. I have updated it since then and adapted it to the situation of the soul talking to the self.

6.   I've heard some people have had conversations with the newly departed via devices like telephones and tape recorders. I've never heard of anyone talking to his or her own soul in this way.

7.   Regions of the brain are associated with speech and understanding of words. However, soul-talk can be understood without any intermediary of words, although at times, certainly, words could be used and these speech areas of the brain (such as Broca's area and Wernicke's area) would show electrical activity during these times.

8.   We do have inner child and parental voices. However, these voices are not soul-talk. See Stephen Wolinsky, *The Dark Side of the Inner Child* (Norfolk, CT: Bramble Books, 1993), p. 123.

9.   Herbert Van Erkelens, "Wolfgang Pauli and the Spirit of Matter," in Pakka Lahti and Peter Mittelstadt, eds., *Symposium on the Foundations of Modern Physics 1990: Quantum Theory of Measurement and Related Philosophical Problems* (Teaneck, NJ: World Scientific Publishing Co., 1991). I wrote about Pauli's attempts to reconcile himself through Jungian interpretation by bringing psychology and physics together in Wolf, *The Dreaming Universe*, chap. 18.

10.   Pauli in C. G. Jung and W. Pauli, *The Interpretation of Nature*, p. 154.

11.   Pauli's principle is a cornerstone of modern physics. In effect it forbids electrons and all particles of matter possessing half-integral spin quantum numbers, that is, 1/2, 3/2, 5/2, and so on, from existing simultaneously in the same quantum state described by the same quantum wave function. For example, not only can no

two electrons in atoms be in the same place at the same time, they also cannot have the same energy or be in the same orbit or have the same angular momentum. If you attempt to move two electrons around until a violation of the Pauli principle occurs, the probability for the occurrence immediately vanishes.

12.   Nothing in classical physics corresponds to the exclusion principle. Objects in quantum physics can have attributes that do correspond classically, such as momentum and position. Each of those attributes exists both at the classical and at the quantum level.

13.   See I. Bernard Cohen, "Kepler's Celestial Music" in *Project Physics Reader 2: Motion in the Heavens*, the Authorized Interim Version (New York: Holt, Rinehart and Winston, 1968–69), pp. 53–75. These structures were the five regular polyhedra, including the tetrahedron (4 sides are equilateral triangles), cube (6 sides are squares), octahedron (8 sides are equilateral triangles), dodecahedron (12 sides are equilateral pentagons), and icosahedron (20 sides are equilateral triangles). As Kepler saw it, there were six concentric spheres, corresponding to the Copernican system of six planets, concentrically nesting the five perfect solids. Sphere (1)—Saturn's sphere, inscribed a cube. Sphere (2)—Jupiter's sphere, was circumscribed by Saturn's cube and inscribed a tetrahedron. Sphere (3)—Mars' sphere was circumscribed by Jupiter's tetrahedron and inscribed a dodecahedron. Earth's sphere (4) circumscribed by Jupiter's dodecahedron and inscribed an icosahedron. Venus's Sphere (5) circumscribed by the Earth's icosahedron and inscribed an octahedron. Venus's octahedron finally inscribed the sphere of Mercury (6).

14.   I discussed a very different version of the trinity in chapter 7.

15.   Wolf (4), *The Body Quantum*, pp. 263–79.

16.   What I present here as a model would hold equally well for a sphere or for any other closed volume.

17.   Compare this model with the rational number mysticism of the Pythagoreans discussed in chapter 3. Communication with the soul is harmonious when a rational collapse occurs in the same sense as music is harmonious when two plucked strings have lengths with a rational ratio.

18.   Rudolph Steiner, *Egyptian Myths and Mysteries* (Hudson, New York: Anthroposophic Press), 1971.

19.   I discussed this situation in several of my earlier books. See Fred Alan Wolf, (2) *Star Wave: Mind, Consciousness, and Quantum Physics* (New York: Macmillan, 1984); (6) *Parallel Universes: The Search for Other Worlds* (New York: Simon & Schuster, 1989); (8) *The Eagle's Quest: A Physicist's Search for Truth in the Heart of the Shamanic World* (New York: Summit, 1991); and (10) *The Dreaming Universe*. Essentially the explanation is that only waves or particles traveling at light-speed or less can be uniquely time-ordered. Anything else can't.

20.   In chapter 4, I explained the significance of the spelling of sacred words. Each letter of a word is also a word and thus each word has hidden meanings revealed when the letters themselves are spelled out.

21.   Nun has two distinct graphs, one when it appears in the middle of a word and the other when it appears at the end of a word as nun-final. This is more or less the same thing as our use of capital and lowercase letters. As such, nun also has two distinct but related meanings.

CHAPTER 14
Some Soulful and Right Answers

1.   D. E. Harding, *On Having No Head: Zen and the Rediscovery of the Obvious* (New York: Arkana, 1986), p. 59.

2.   See chapter 13.

3.   Although, as I mentioned in chapter 7, an effort is being made to study consciousness scientifically. So, while it still appears science is bogged down in materialism in this effort, the spiritual side of science is undergoing resurrection.

4.   My statistics might be off by some percentage points, but the spirit of my statement is true.

5.   Today physicists are investigating string theory, a model that looks at fundamental particles as vibrations of subatomic invisible strings. Perhaps particles are strings of spatially reflected quantum waves appearing as strings on tiny spatial scales. This would certainly fit my model.

6.   In his film *The Purple Rose of Cairo*, Woody Allen used the technique of a character from a movie walking out of the screen and interacting with a real person. Of course, it was only a movie.

Bibliography

Aharonov, Y., and M. Vardi. "Meaning of an Individual 'Feynman Path'." *Physical Review Digest,* 21, no. 8, (April 15, 1980), pp. 2235–40.

Albert, David Z. *Quantum Mechanics and Experience.* Cambridge: Harvard University Press, 1992.

——— (2). "On Quantum-Mechanical Automata." *Physics Letters,* 98A, nos. 5, 6 (October 24, 1983), pp. 249–52.

Alston, William P., ed. *Religious Belief and Philosophical Thought.* New York: Harcourt Brace Jovanovich, 1963.

"Angels, Aliens, and Archetypes." *Revision,* 11, nos. 3, 4, parts 1, 2.

Armstrong, Karen. *A History of God: The 4,000-Year Quest of Judaism, Christianity, and Islam.* New York: Alfred A. Knopf, 1994.

Avens, Roberts. *Imaginal Body: Para-Jungian Reflections on Soul, Imagination and Death.* Washington, D. C.: University Press of America, 1982.

Bamford, Christopher, ed. *Homage to Pythagoras: Rediscovering Sacred Science.* Hudson, NY: Lindesfarne Press, 1994.

Barrett, William. *Death of the Soul: From Descartes to the Computer.* Garden City: Anchor Press, 1986.

Bass, Ludvik. "The Mind of Wigner's Friend." *Hermathena: A Dublin University Review,* no. 112, (1971) pp. 52–68.

The Bhagavad Gita: The Gospel of the Lord Khrishna. Translated by Shri Purohit Swami. New York: Vintage, 1977.

Bohm, David. *Wholeness and the Implicate Order.* Boston: Routledge and Kegan Paul, 1980.

Brinkley, Dannion (with Paul Perry). *Saved by the Light: The True Story of a Man Who Died Twice and the Profound Revelations He Received.* New York: Villard, 1994.

Browne, Malcolm W. "The Soul Weighs In: Scientist Says He Can Measure Our Spiritual Selves." *San Francisco Chronicle: Sunday Punch* taken from the *New York Times,* January 16, 1994.

Budge, E. A. Wallis. *The Book of the Dead.* London: Arkana, 1994.

——— (2). *Egyptian Religion: Egyptian Ideas of the Future Life.* London: Arkana, 1987.

——— (3). *Osiris and The Egyptian Resurrection.* New York: Dover, 1973.

——— (4). *Egyptian Magic.* New York: Dover, 1971.

——— (5). *The Gods of the Egyptians.* New York: Dover, 1969.

Buksbazen, John Daishin. *To Forget the Self,* vol. 3, Zen Writings Series. Los Angeles: Zen Center of Los Angeles, n. d.

Byrom, Thomas. *Dhammapada: The Sayings of the Buddha.* Boston: Shambhala, 1976.

Capra, Fritjof. *The Tao of Physics.* Boston: Shambhala, 1975.

Capra, Fritjof and David Steindl-Rast. *Belonging to the Universe: Explorations on the Frontiers of Science and Spirituality.* San Francisco: HarperSanFrancisco, 1991.

Cauldron Productions. *Evil: The Cosmic Shadow.* New York: House Publications Service, 1994.

Chalmers, David J. "The Puzzle of Conscious Experience." *Scientific American* (December 1995), pp. 80–86.

Charon, Jean. *The Unknown Spirit.* London: Coventure, 1983.

Cohen, I. Bernard. "Kepler's Celestial Music." In *Project Physics Reader 2: Motion in the Heavens.* The Authorized Interim Version, pp. 53–75. New York: Holt, Rinehart and Winston, 1968–69.

Corbin, Henri. *Mundis Imaginalis or the Imaginal and the Imaginary.* Ipswich, England: Golgonooza Press, 1976.

Cott, Jonathan. *Isis and Osiris: Exploring the Goddess Myth.* New York: Doubleday, 1994.

Cousineau, Phil, ed. *Soul: An Archaeology. Readings from Socrates to Ray Charles.* San Francisco: HarperSanFrancisco, 1994.

————— (2). *The Soul of the World.* San Francisco: HarperSanFrancisco, 1993.

Crawford, Charles, ed. and trans. *A Mirror for Simple Souls: The Mystical Work of Marguerite Porete.* New York: Crossroads, 1990.

Crick, Francis. *The Astonishing Hypothesis: The Scientific Search for the Soul.* New York: Charles Scribner's Sons, 1994.

David-Neel, Alexandra. *Buddhism: Its Doctrines and Its Methods.* New York: Avon, St. Martin's Press, 1979.

David-Neel, Alexandra and Lama Yongden. *The Secret Oral Teachings in Tibetan Buddhist Sects.* San Francisco: City Lights Books, 1967.

Davidson, John. *The Secret of the Creative Vacuum: Man and the Energy Dance.* Essex, England: C. W. Daniel, 1989.

Davies, Paul. *The Mind of God: The Scientific Basis for a Rational World.* New York: Touchstone, 1993.

Devereux, Paul, John Steele and David Kubrin. *Earthmind—Is the Earth Alive?* New York: Harper and Row, 1989.

Dirac, P. A. M. "The Early Years of Relativity." In *Albert Einstein: Historical and Cultural Perspectives. The Centennial Symposium in Jerusalem.* Edited by Gerald Holton and Yehuda Elkana. Princeton: Princeton University Press, 1982.

Dossey, Larry, M.D. *Recovering The Soul: A Scientific and Spiritual Search.* New York: Bantam, 1989.

Douglas-Klotz, Neil. *Desert Wisdom: Sacred Middle Eastern Writings from the Goddess through the Sufis, Translations and Commentary.* San Francisco: HarperSanFrancisco, 1995.

Dunne, J. W. *The New Immortality.* London: Faber and Faber, 1938.

Easwaran, Eknath. *Dialogue with Death: The Spiritual Psychology of the Katha Upanishad.* Berkeley: The Blue Mountain Center of Meditation, 1981.

Einstein, Albert. *Ideas and Opinions.* New York: Crown, 1954.

Ellis, Normandi. *Awakening Osiris: A New Translation of The Egyptian Book of the Dead.* Grand Rapids: Phanes Publishing, 1988.

Ellis, William. *The Idea of the Soul: In Western Philosophy and Science.* London: George Allen and Unwin, 1940.

Epstein, Mark. "Are We All Hungry Ghosts?" *Inquiring Mind,* 11, no. 2 (spring 1995), p. 11.

Erkelens, Herbert Van. "Wolfgang Pauli and the Spirit of Matter." In Pakka Lahti and Peter Mittelstadt, eds., *Symposium on the Foundations of Modern Physics 1990: Quantum Theory of Measurement and Related Philosophical Problems.* Teaneck, NJ: World Scientific Publishing, 1991.

Fadiman, James, and Robert Frager. *Personality And Personal Growth.* New York: Harper and Row, 1976.

Faulkner, Raymond, trans. *The Egyptian Book of the Dead: The Book of Going Forth by Day.* San Francisco: Chronicle Books, 1994.

Feuerstein, Georg. "The Changing Fortunes of the Soul." *Magical Blend: Unveiling Body and Soul,* 37 (January 1993), p. 11.

Feynman, R. P. *The Theory of Fundamental Processes.* Reading, MA: The Benjamin/Cummings Publishing, 1962–82.

———— (2). *QED: The Strange Theory of Light and Matter.* Princeton: Princeton University Press, 1985.

———— (3). "The Value of Science." *Project Physics Reader 1: An Introduction to Physics, Concepts of Motion,* pp. 1–8. Authorized Interim Version. New York: Holt, Rinehart and Winston, 1968–69.

Feynman, Richard P. and Steven Weinberg. *Elementary Particles and the Laws of Physics: The 1986 Dirac Memorial Lectures.* New York: Cambridge University Press, 1987.

Fordham, Frieda. *An Introduction to Jungian Psychology.* New York: Penguin, 1953.

Frankfort, Henri, Mrs. H. A. Frankfort, John A. Wilson, and Thorkild Jacobsen. *Before Philosophy: The Intellectual Adventure of Ancient Man. An Essay on Speculative Thought in the Ancient Near East.* Baltimore: Penguin, 1949.

Friedlander, Shems. *The Whirling Dervishes.* New York: SUNY, 1992.

Gold, E. J. *American Book of the Dead.* San Francisco: AND/OR Press, 1975.

Goswami, Amit. *The Self-Aware Universe: How Consciousness Creates the Material World.* New York: Tarcher/Putnam, 1993.

Grinberg-Zylberbaum, J., M. Delaflor, L. Attie, and A. Goswami. "The Einstein-Podolsky-Rosen Paradox in the Brain: The Transferred Potential." *Physics Essays,* 7, no. 4 (1994), pp. 422–28.

Grof, Stanislav. *Beyond the Brain: Birth, Death, and Transcendence in Psychotherapy.* Albany: State University of New York, 1985.

Grof, Stanislav, M.D., with Hal Zina Bennett. *The Holotropic Mind.* San Francisco: HarperCollins, 1992.

Grosso, Mike. *Soulmaker: True Stories from the Far Side of the Psyche.* Norfolk: Hampton Roads, 1992.

———— (2). "The Status of Survival Research: Evidence, Problems, Paradigms." *Noetic Sciences Review,* 32 (1994), pp. 12–20.

Grun, Bernard. *The Timetables of History: A Horizontal Linkage of People and Events.* New York: Touchstone, 1982.

Harding, D. E. *On Having No Head: Zen and the Rediscovery of the Obvious.* New York: Arkana, 1986.

Hayward, Jeremy, and Francisco Varela. *Gentle Bridges: Conversations with the Dalai Lama on the Sciences of Mind.* Boston: Shambhala, 1992.

Heinze, Ruth-Inge. *Tham Khwan—How to Contain the Essence of Life—A Socio-Psychological Comparison of a Thai Custom.* Singapore: Singapore University Press, 1982.

Highwater, Jamake. *The Primal Mind: Visions and Reality in Indian America.* New York: Harper and Row, 1981.

Hillman, James, and Michael Ventura. *A Hundred Years of Psychotherapy and the World's Getting Worse.* San Francisco: HarperSanFrancisco, 1992.

Holland, Peter. *The Quantum Theory of Motion.* Boston: Cambridge University Press, 1995.

Hopkins, Thomas J. *The Hindu Religious Tradition.* Belmont, CA: Wadsworth, 1971.

Illion, Theodore. *Darkness Over Tibet.* London: Rider, 1937; Stelle, IL: Adventures Unlimited Press, 1991.

———— (2). *In Secret Tibet.* London: Rider, 1937; Stelle, IL: Adventures Unlimited Press, 1991.

Johnson, Robert A. *We: Understanding the Psychology of Romantic Love.* San Francisco: HarperSanFrancisco, 1983.

Jung, C. G. and W. Pauli. *The Interpretation of Nature and the Psyche. Synchronicity: An Acausal Connecting Principle* (Jung) and *The Influence of Archetypal Ideas on the Scientific Theories of Kepler* (Pauli). New York: Bollingen Foundation, Pantheon, 1955.

Kerényi, Karl. *Hermes: Guide of Souls.* Translated by Murray Stein. Dallas: Spring Publications, 1992.

Killheffer, Robert K. "The Consciousness Wars." *Omni* (October 1993), p. 50.

Kline, Morris. *Mathematics: The Loss of Certainty.* New York: Oxford University Press, 1980.

Lamy, Lucie. *Egyptian Mysteries.* New York: Thames and Hudson, 1981, 1989.

Mailer, Norman. *Ancient Evenings.* Boston: Little, Brown, 1983.

Maritain, Jacques. *The Range of Reason.* New York: Charles Scribner's Sons, 1952. See also: "A Proof of the Immortality of the Soul." In *Religious Belief and Philosophical Thought.* Edited by William P. Alston. New York: Harcourt Brace Jovanovich, 1963.

Masters, Robert. *The Goddess Sekhmet: Psychospiritual Exercises of the Fifth Way.* St. Paul: Llewellyn, 1990.

McKeon, Richard. *Introduction to Aristotle:* 2d ed. rev. and enl. Chicago: University of Chicago Press, 1973.

Misner, Charles W., Kip S. Thorne, and John Archibald Wheeler. *Gravitation.* San Francisco: W. H. Freeman, 1973.

Mittleman, M. H., and Fred A. Wolf, "Coherent Scattering of Photons by Atomic Hydrogen," *Physical Review,* 128 (1962), p. 2686.

Modi, Dr. Bhupendra Kumar. *Hinduism: The Universal Truth.* New Delhi: Brijbasi Printers in conjunction with the World Buddhist Cultural Foundation, New Delhi, 1993.

Moore, Thomas. *Care of the Soul: A Guide for Cultivating Depth and Sacredness in Everyday Life.* New York: HarperCollins, 1992.

Morick, Harold, ed. *Introduction to the Philosophy Of Mind: Readings from Descartes to Strawson.* Glenview: Scott, Foresman, 1970.

Mouravieff, Boris. *Gnosis: Study and Commentaries on the Esoteric Tradition of Eastern Orthodoxy.* Book 1, *Exoteric Cycle.* Translated by S. M. Wissa. Newbury, MA: Agora Books/Praxis Institute Press, 1989.

Murphy. Michael. *The Future of the Body: Explorations Into the Further Evolution of Human Nature.* Los Angeles: Jeremy P. Tarcher, 1992.

Parrinder, Geoffrey. *The Indestructible Soul: The Nature of Man and Life After Death in Indian Thought.* London: George Allen and Unwin, 1973.

Penrose, Roger. *Shadows of the Mind: A Search for the Missing Science of Consciousness.* New York: Oxford University Press, 1994.

Peursen, C. A. van. *Body, Soul, Spirit: A Survey of the Body-Mind Problem.* Translated by Hubert H. Hoskins. London: Oxford University Press, 1966.

Plato. *The Last Days of Socrates.* Translated by Hugh Tredennick and Harold Tarrant. New York: Penguin Classics, 1993.

———— (2). *The Republic.* Revised ed. Translated by Desmond Lee. New York: Penguin Classics, 1987.

———— (3). *Theaetetus.* Translated with an essay by Robin Waterfield. New York: Penguin Classics. 1987.

———— (4). *Timaeus and Critias.* Translated with an Introduction by Desmond Lee. New York: Penguin Classics, 1971.

Podolny, R. *Something Called Nothing: Physical Vacuum: What Is It?* Translated by Nicholas Weinstein. Moscow: Mir Publishing, 1986.

Porush, David. "Finding God in the Three-Pound Universe: The Neuroscience of Transcendence," *Omni* (October 1993), p. 60.

Rahula, Walpola. *What the Buddha Taught.* 2nd ed. New York: Grove Press, 1974.

Renou, Louis. *Religions of Ancient India.* New York: Schocken Books, 1968.

Ross, Nancy Wilson. *Buddhism: A Way of Life and Thought.* New York: Alfred A. Knopf, 1980.

Russell, Bertrand. *History of Western Philosophy and its Connection with Political and Social Circumstances from the Earliest Times to the Present Day.* London: George Allen and Unwin, 1961.

Schiff, Leonard I. *Quantum Mechanics.* 3d ed. New York: McGraw-Hill, 1968.

Schrödinger, Erwin. *My View of the World.* Originally published in German. Hamburg-Vienna: Paul Zsolnay Verlag GMBH, 1961. English edition. Cambridge, England: Cambridge University Press, 1964. Reprint, Woodbridge, CT: Ox Bow Press, 1983.

———— (2). *What is Life?* and *Mind and Matter.* Cambridge, England : Cambridge University Press, 1967.

Schueler, Gerald and Betty. *Coming Into the Light: Rituals of Egyptian Magick.* St. Paul: Llewellyn, 1989.

Smith, Huston. *The Religions of Man.* New York: Harper and Row, 1958.

Smoley, Richard. "My mind plays tricks on me." *Gnosis Magazine* (spring 1991), p. 12.

Sperling, G. "Information Available in Brief Visual Presentations." *Psychological Monographs* 74, no. 11 (1960).

Steiner, Rudolph. *Egyptian Myths and Mysteries.* Hudson, New York: Anthroposophic Press, 1971.

———— (2). *Metamorphoses of the Soul.* Vol. 2, 2d ed. London: Anthroposophic Press, 1990.

Suarès, Carlo. *The Cipher of Genesis: The Original Code of the Qabala as Applied to the Scriptures.* Berkeley: Shambhala, 1970.

———— (2). *Les Spectrogrammes de l'Alphabet Hebraïque.* Geneva: Mont-Blanc, 1973.

———— (3). "The Cipher of Genesis," *Tree 2: Yetzirah.* From a lecture by Suarès. Reprint from *Systematics* 8, no. 2 (September 1970), edited by David Meltzer. Santa Barbara: Christopher Books, 1971.

Swineburn, Richard. *The Evolution of the Soul.* Oxford, England: Clarendon Press, 1986.

Talbot, Michael. *The Holographic Universe.* New York: HarperCollins, 1991.

Thakkar, Hirabhai. *Theory of Karma.* Amreli, India: Shri Rameshwer Printing, 1988.

Tilby, Angela. *Soul: God, Self and The New Cosmology.* New York: Doubleday, 1992.

Tipler, Frank J. *The Physics of Immortality: Modern Cosmology, God and the Resurrection of the Dead.* New York: Doubleday, 1994.

Tulku, Tarthang. *Knowledge of Time and Space.* Berkeley: Dharma Publishing, 1990.

Vivekananda, Swami. *Jnana-Yoga.* New York: Ramakrishna-Vivekananda Center, 1982.

von Franz, Marie-Louise. *Number and Time: Reflections Leading to a Unification of Depth Psychology and Physics.* Evanston: Northwestern University Press, 1974.

Walker, Barbara G. *The Woman's Encyclopedia of Myths and Secrets.* New York: Harper and Row, 1983.

Weinberg, Stephen. *The First Three Minutes.* New York: Basic Books, 1977.

Weiner, Norbert. *God and Golem, Inc.: A Comment on Certain Points where Cybernetics Impinges on Religion.* Cambridge: MIT Press, 1964.

Weiss, Brian L., M.D. *Many Lives, Many Masters: The True Story of a Prominent Psychiatrist, His Young Patient, and the Past-Life Therapy that Changed Both Their Lives.* New York: Simon and Schuster, 1988.

West, John Anthony. *Serpent in the Sky: The High Wisdom of Ancient Egypt*. Wheaton, IL: Theosophical Publishing House, 1993.

Wilson, A. N. *Jesus: A Life*. New York: Ballantine, 1993.

Wilson, Ian. *The After Death Experience: The Physics of the Non-Physical*. New York: William Morrow, 1987.

Wolf, Fred Alan. *Taking the Quantum Leap: The New Physics for Nonscientists*. San Francisco: rev. ed. New York: HarperCollins, 1989.

———— (2). *Star Wave: Mind, Consciousness, and Quantum Physics*. New York: Macmillan, 1984.

———— (3). "The Quantum Physics of Consciousness: Towards a New Psychology." *Integrative Psychology* 3 (1985), pp. 236–47.

———— (4). *The Body Quantum: The New Physics of Body, Mind, and Health*. New York: Macmillan, 1986.

———— (5). "The Physics of Dream Consciousness: Is the Lucid Dream a Parallel Universe?" *Lucidity Letter* 6, no. 2 (December 1987), pp. 130–35.

———— (6). *Parallel Universes: The Search for Other Worlds*. New York: Simon and Schuster, 1989.

———— (7). "On the Quantum Physical Theory of Subjective Antedating." *Journal of Theoretical Biology* 136 (1989), pp. 13–19.

———— (8). *The Eagle's Quest: A Physicist's Search for Truth in the Heart of the Shamanic World*. New York: Summit, 1991.

———— (9). "The Dreaming Universe." *Gnosis* 22 (winter 1992), pp. 30–35.

———— (10). *The Dreaming Universe: A Mind-Expanding Journey into the Realm Where Psyche and Physics Meet*. New York: Simon and Schuster, 1994. Reprint, Touchstone, 1995.

———— (11). "The Body in Mind." *Psychological Perspectives: A Journal of Global Consciousness Integrating Psyche, Soul and Nature* 30 (fall-winter 1994), pp. 22–35.

———— (12). "The Quantum Mechanics of Dreams and the Emergence of Self-Awareness." In *Toward a Scientific Basis for Consciousness*. Edited by S. R. Hameroff, A. W. Kaszniak, and A. C. Scott. Boston: MIT Press, 1996.

Wolinsky, Stephen. *The Dark Side of the Inner Child*. Norfolk, CT: Bramble Books, 1993.

Zaleski, Carol. *Otherworld Journeys: Accounts of Near-Death Experiences in Medieval and Modern Times*. New York: Oxford University Press, 1987.

Zohar, Danah. *The Quantum Self: Human Nature and Consciousness Defined by the New Physics*. New York: Quill/William Morrow, 1990.

Index